Finite Element Modeling of Textiles in Abaqus™ CAE

Finite Element Modeling of Textiles in Abaqus™ CAE

Izabela Ciesielska-Wrobel

CRC Press
Taylor & Francis Group
Boca Raton London New York

CRC Press is an imprint of the
Taylor & Francis Group, an **informa** business

CRC Press
Taylor & Francis Group
52 Vanderbilt Avenue,
New York, NY 10017

Printed on acid-free paper

International Standard Book Number-13: 978-1-498-75373-9 (Hardback)

Visit the Taylor & Francis Web site at
http://www.taylorandfrancis.com

and the CRC Press Web site at
http://www.crcpress.com

MIX
Paper from responsible sources
FSC
www.fsc.org
FSC® C013056

Printed and bound in Great Britain by
TJ International Ltd, Padstow, Cornwall

To my husband, Jakub, my daughter, Abigail, and to my mother, Eleonora, without whom this book would not have been completed—thank you for the encouragement.

Disclaimer

Although the images in this book are black and white, the description of each of them and the matters discussed in this book refer to colors of the models available while performing modeling.

Contents

Preface

Fortunately, there are plenty of interesting books concerning Finite Element Modeling (FEM) techniques; therefore, as a reader, you are not completely abandoned and lonely in the scientific world of modeling. Usually, feeling lost starts when you are faced with real modeling problems, considering a very specific, unique matter strictly related to your field. The problem is that there are usually very few people dealing with similar aspects at any given point in time. Thus, most of the necessary information needs to be mined out of a great number of pieces of scientific elaborations in the form of scientific papers or monographs. Guides, on the other hand, could be a great source of detailed and specific information; but, due to many reasons, their existence on the book market is limited. The crucial reason is the fact that almost no one wants to reveal their geometry creation and modeling techniques. In order to obtain useful information on this subject, it is suggested to follow the guidelines from this book and follow workshops on step-by-step actions that are needed to create geometry of objects and perform the modeling process.

This book is the first elaboration that exists on the book market that deals with performing the modeling process of selected textile objects and human skin in Abaqus™ CAE (Complete Abaqus Environment) software. It reveals the secrets of modeling of textile objects in a reliable and reasonable way—namely, introducing more advanced features gradually—so that the reader is able to understand the methodology of building the models and their specific nature.

The application of Abaqus CAE software with numerical tools was reserved in the past to solve complex mechanical equations. It has changed lately and FEM is widely applied in materials science, heat transfer phenomena, and by civil engineers, biomechanical engineers, etc. Application of FEM in the areas of textiles and haptic perception of textiles not only helps in better understanding of textiles themselves, but also their interaction with human skin: known as haptic perception of textiles—sometimes called tactile perception—this is the feeling of textiles against the skin, or simply "hand" of textiles.

Textiles are very difficult objects to model, as they are uneven in most cases, so a great degree of idealization is necessary to approach their modeling. Similarly, human tissue is not a homogeneous material. The structure and physical properties of both textiles and skin are described in detail so that the nature of both objects—textile and skin—can be understood, and their models can be created and processed with understanding and caution when it comes to interpretation.

This book is intended for both beginners in the field of FEM of textiles as well as those who have some knowledge, even for those that are quite advanced in the area. The reason that this book complies with the requirements of many readers—representing different levels of knowledge—is the fact that combining very specific information related to modeling of textile structures using lay language (in some cases) allows readers to understand the nature of some phenomena. Naturally, a brief scientific review of the modeled objects will precede each introduced topic to provide the reader with reliable information so that he/she can also gain encyclopedic knowledge related to both the object and its physical/mechanical characteristics. Materials provided in this book can be utilized not only as a self-study guide, but also as class materials for university courses concerning FEM and textile modeling.

The author is grateful to the European Commission, the Seventh European Framework, Marie Curie International Outgoing Fellowship for financing "Modeling of Human Body and Protective Textiles for Estimation of Skin Sensorial Comfort and Life Risk of Fire Fighters Working in Extreme External Conditions" (known as the Magnum Bonum project, no. 622043), which has made the writing and editing of this book possible.

Izabela Ciesielska-Wrobel, PhD, Eng.
Ghent University, Belgium
June 2018

Acknowledgments

The author is grateful to the European Union, Seventh European Framework for funding the Magnum Bonum Project (grant number 622043) within the Marie Curie International Outgoing Fellowship that partially allowed preparation of this book. Many thanks to my former students, Nina Bollaert, Laura Dens, and Wouter Dujardin from Belgium, for their help in editing Section 4.2 Model of Skin Touching an Object, and to Ranjitprakash Sundaramurthi, a graduate of the Ohio State University in the United States, for his help in the analysis of a knit structure model's geometry.

Author

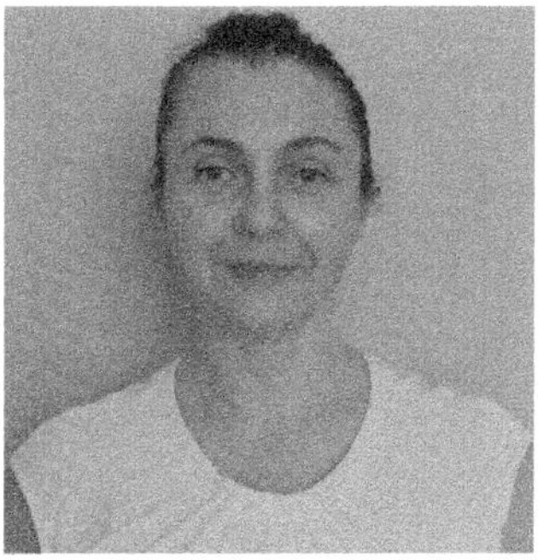

Prof. Izabela Ciesielska-Wrobel is a postdoctoral senior researcher at Ghent University, Department of Materials, Textiles and Chemical Engineering, and was a visiting scholar at North Carolina State University, College of Textiles, Textile Performance and Comfort Center (known as TPACC). She obtained her master's degree in Textile Engineering at Lodz University of Technology in Poland, where, after completing her PhD studies and PhD defense in 2007, she continued her studies on comfort in textiles, including protective textiles as well as haptic perception of textiles, as an assistant professor and adjunct. She has published more than 50 articles, including refereed papers and conference materials. She is a coauthor of two patents. She is a beneficiary of two prestigious and competitive European grants, namely the Marie Curie Intra-European Fellowship for Career Development (2010 – 2012) and the Marie Curie International Outgoing Fellowship (2014 – 2017). Currently, she lives in North Carolina, USA.

1

Introduction to Creation of Geometry and Modeling with Abaqus CAE

1.1 Introduction

The aim of the book is to provide textile engineers with a practical guide to finite element modeling (FEM) in Abaqus CAE (Complete Abaqus Environment) software. The guide is in the form of step-by-step procedures concerning fiber, yarn, knitted fabric modeling, as well as their contact with skin so that the simulation of haptic perception between textiles and skin can be provided. The specific modeling procedure will be preceded by a theoretical background concerning the mechanical characteristics of modeled elements or phenomena (e.g., in the case of the modeling of haptic perception of textiles, a detailed model of the soft tissue (human skin) and modeled textile objects will be given). Models will be validated and discussed. In addition, virtual object test results will be presented and compared with the outcome of the modeling process. The discussion of the convergences and divergences between real case studies and their models will be analyzed.

The scope of the book is as follows:

1. Theoretical background concerning mechanical characteristics and structure of basic textile products.
2. FEM of selected objects, namely single fiber, three-ply yarn, plain stitch knitted fabric, single-layer skin model, three-layer skin model (skin compliance), and contact of the skin with fabric (skin compliance, friction).
3. The results of the physical tests on textiles will be provided to verify the proposed modeling approach.

The process of modeling is time consuming, especially for beginners in this field or for those who have knowledge of the subject but are not familiar with the software that allows modeling. Normally, one needs to follow the basic instructions provided by the developer of the software (Abaqus CAE User's Manual). To create a complex, original model, one needs to collect information from many existing sources and apply a method of trial and error when using the collected information. In many cases, instead of modeling existing phenomena, one creates an animation of phenomena without applying careful attention to the simulation of phenomena. This book will guide users of finite element software (not only Abaqus CAE users, as the modeling rules are common for most of the existing FEM software) from the stage of complex graphic creation to material selection, mesh, contact properties, and boundary conditions up to analysis of the results (Dassault Systèmes, 2016).

In everyday life, one needs to perform plenty of physical tests in the lab to make sure that the specific fabric complies with required standards or simply to verify the quality of materials. In the case of scientific studies, the cost of operator-tested samples is high. That is why it is sometimes easier to perform modeling on the computer instead of carrying out laboratory tests. The other great example of the application of FEM using software is the fact that very complex and expensive samples (e.g., carbon composite materials) can only be tested a few times. The introduction of modifications after each test is extremely costly. This is the moment when FEM and Abaqus CAE come into play. This method certainly contributes to a reduction of tests and development cycling times. Eventually, it does induce quicker turnaround times and saves money.

1.1.1 Textiles—Why Model Them?

Textiles are unique products; they can be made of different raw materials, starting from natural and ending with highly modified synthetics. Their composition, structure, and abilities determine their final application. In order to predict whether a specific raw material or product can be successfully applied to a complex structure—for example, carbon fabric composite application used in car hulls—one usually conducts a tedious time-, energy-, and financial resources-consuming process. This process is aimed at verifying whether this specific compound of materials with a specific architecture meets the requirements, or whether it fails before achieving the desired requirement—for example, a high value of breaking force. If this specific compound fails, one needs to perform another verification test, and then another, modifying the compound to achieve the goal, which in this case could be meeting the mechanical criteria—for example, achieving a tensile strength of 110 MPa.

Nowadays, to avoid a trial-and-error method becoming a very costly approach, one utilizes a high-performance computational software that is able to provide the user with an extremely precise and fast solution. Unfortunately, the word "fast" is a key keyword here. Why? Well, to perform any verification or modeling of the objects using this type of software, one needs to gain an understanding of the phenomena (e.g., breaking force), objects (e.g., fiber, yarns), and the tool itself (software, e.g., Abaqus). These are key issues related to the FEM of all modeled objects and their materials, not only textiles (although textiles are the most important subjects in this book), which all Abaqus users need to deal with.

Most textiles can be described as uneven, which makes them truly unique objects, but unfortunately in many cases, this special feature is omitted, simplifying the model objects.

In the context of this book, the process of modeling the textiles is understood as creating the geometry of the textile object, ascribing to it the necessary material parameters, combining them into one assembly, preparing a step-by-step scenario for the assembled objects, selecting the parameters to be calculated by the software (the final values, the outcome of the modeled phenomenon), establishing the forces, loads working on the object in the system and boundary conditions of the system or its elements, performing the analytical part, assessing the results, and finally reporting.

This kind of procedure will be performed for several textile objects, and at the end, the examples of the real study utilizing the modeling of textiles will be presented.

The modeling of textiles is not only conducted to save time and financial resources, which are usually spent when performing tests on textiles in a traditional manner in the metrology laboratories, but also to gain a better understanding of the phenomena taking place in the textiles assembled or for educational purposes when one can observe interactions taking place between twisted yarns.

1.1.2 Finite Element Analysis and Finite Element Models

The finite element analysis (FEA) method was introduced by M.J. Turner et al. (1956) in the article titled "Stiffness and Deflection Analysis of Complex Structures," where a method for yielding accurate structural data allowing further dynamic analysis was introduced. The concept of FEA has been widely applied in different fields of engineering and bioengineering. It is usually defined as a discretization technique in mechanics, specifically in structural mechanics. As the term suggests, the technique requires dividing the model (object) into elements that have less complicated geometry than the geometry of the whole initial model. This simplification of the geometry allows elements to be obtained (fragments of the initial model). These are called finite elements. Any analytical attempt to consider the whole model, without it being subdivided, could be extremely difficult or even impossible to perform. The division, called discretization, helps to investigate the model through its elements. The mathematical answer regarding the whole model can be considered as an estimate of all the elements taken into consideration collectively. The idea is to gain a better understanding of all the elements separately so that one can understand the whole object and predict its behavior. The purpose of this is to predict the response of a model to some form of external loading, stress, external elevated temperature or to some nonequilibrium initial conditions.

The finite element method in mathematics is a specific mathematical technique (approach) utilized to approximate some solutions to problems presented in the form of differential equations. In many cases, finite element method refers to finite element models; however, in these cases, it is specifically mentioned that one is referring to the modeled objects, not the method that was applied to provide the solution.

1.1.3 Abaqus CAE and Other Software as Tools for Modeling

Commercially available FEA software gained great popularity among engineers, especially constructional engineers in industry, architecture, tectonics, and among researchers representing different fields (e.g., bioengineers in academia). Despite the fact that this book is meant for textile engineers and all the examples presented in the book refer to textiles, the methodology of creating a model, no matter what the object of the modeling is, remains the same and can be easily applied in different fields.

The idea of applying FEA software to model an object or phenomena existing in real life or only hypothetically, appeared when it was found that performing a trial-and-error testing method became too expensive and too robust to continue. Why prepare the object, test it and analyze the results only to find out that it is a failure when the same action can be taken using a powerful computational technique to obtain approximate testing results or solutions to a variety of engineering problems without leaving the desk and the computer?

The idea of modeling in Abaqus CAE or any other FEA software listed in this book is to recreate an environment (load, boundary conditions) that surrounds an object with a specific geometry (model) and to imitate the real-life circumstances happening to this model (torque or external pressure).

Imagine working on a textile sample in the lab. Your textile object is prepared and then destroyed in the course of a test (e.g., a tearing test). There is at least one thing you know and at least one thing you do not know about this sample. You definitely know that your sample was damaged in the course of the test when you applied a specific force to tear the sample apart. What you do not know is exactly in which place the sample was destroyed and how did it started and ended, in which spots on the sample. In order to have a better insight into a sample, a microscopic imaging can be applied. This may answer some of your questions, but it will probably lead you back to the lab and you will be forced to perform more tests. This is not an issue if your samples are not expensive in terms of raw materials and the methodology used in preparing the sample for the test. Now, imagine a model with the same geometry as your sample. This model was created in the FEA software, where you initiate tearing of the sample using a relatively simple combination of the conditions in the software (steps, contact, load, boundary conditions). As soon as the model is created, you can initiate the destruction of samples, investigating the selected moment and places on the sample.

You are also free to change the applied forces to verify whether different loading exerts the same or different effect on the tested (modeled) sample. In fact, you may decompose the whole model, observing only selected elements in the model. This ability of FEA software that allows discretizing the whole model and observing selected elements or points (nodes) makes the software a useful e-tool for solutions to engineering problems.

Released in 1978, Abaqus CAE is an environment that has graphical applications. In 2005, it was purchased by Dassault Systèmes and has been cocreating the SIMULIA brand. This allows interactive operating with the designed objects. It also allows the rapid and relatively easy creation of models of objects that have a simple geometry. In the case of a complex model, the software allows the geometry to be imported; and, by producing or importing the geometry of the structure, and decomposing it into smaller fragments, it can be easily analyzed. This operation is called meshing. If the model is checked and submitted, Abaqus CAE can monitor and control the process of analysis (so-called "job"). The visualization module can then be used to interpret the results.

There is a number of Abaqus platforms that can be utilized depending on the needs of the user. The first one, and the most common, is Abaqus/Standard, which is a general-purpose analysis product that can provide solutions to a great variety of linear and nonlinear problems involving the static, dynamic, thermal, and electrical response of modeled elements. The examples presented in this guide refer to this specific platform. Abaqus/Standard solves a system of equations implicitly at each solution "increment." On the other pole, one may find that Abaqus/Explicit can provide a solution over time in small time increments without solving a coupled system of equations at each increment. This is suitable for modeling brief, transient, dynamic events, such as impact and blast problems, or the change of the temperature on one surface related to the heating of another surface in close vicinity to the first one. Abaqus/Explicit is not utilized in this guide.

Thermal convective flow and incompressible flow problems may be solved by computational fluid dynamics (CFD) tool, Abaqus/CFD; however, this does not feature in this guide.

Abaqus CAE software has its direct and indirect competitors, which means that some of them can provide the same platform for the solution of specific problems, while some of them have different profiles and are more focused on specific areas (e.g., biomechanics).

The existing list of software allowing FEM of objects and phenomena is very long, and the list presented here contains only some selected multipurpose software, the most popular in the opinion of the author and/or useful in terms of their application in the area of skin and textile modeling. It does not mean that other existing software is worse. It just means that the author was not aware that they exist or that they are excellent but are meant for applications that are irrelevant in here (e.g., sets of PLAXIS 2D and PLAXIS 3D developed and marketed by Plaxis BV), which are specially designed for soil and rock mechanics with the emphasis on the analysis, design, and simulation of underground constructions and the soil-structure interaction (http://plaxis.nl/); Bridge, Lusas Bridge LT, and Lusas Bridge Plus by LUSAS is a set of software graded from a basic version to a sophisticated one, where one may perform an FEA of all types of bridge structures. Bridge LT is suitable for linear static analysis of structural frames and grillages. Bridge is suitable for everyday linear static and linear dynamic analysis using beams, shells, solids, and joints. Bridge Plus includes an extended advanced high-performance element library which allows for more advanced analyses to be undertaken. Examples of real-life industry use of LUSAS software products are dismantling of the San Francisco-Oakland Bay Bridge East Main Span, Stockholm Waterfront Congress Centre, and analysis of composite components (http://lusas.com/index.html).

The list created by the author contains the following:

1. Automatic Dynamic Incremental Nonlinear Analysis (ADINA) by Adina R&D, Inc. is a comprehensive software that offers a wide range of capabilities based on reliable and efficient finite element procedures. It can be applied when analyzing structural, fluid, heat transfer, electromagnetic, and multiphysics problems, including fluid–structure interaction and thermomechanical coupling (http://adina.com/adina-structures.shtml).
2. Simulation Mechanical by Autodesk is an FEM and FEA software featuring an integrated Autodesk Nastran FEA solver software that provides designers, engineers, and analysts with fast and flexible tools for finite element analysis and modeling. Simulation Mechanical is only one out of a very long list of different software provided by Autodesk (http://autodesk.com/products/simulation-mechanical/overview).
3. HyperWorks by Altair is a full Computer-Aided Engineering (CAE) program suite (including pre/postprocessor, solvers, optimization solver, and parameter study solver) that offers many new functionalities in such areas as lightweight design, composites structure analysis, including textile-like structures, efficient modeling and meshing, multiphysics and multidisciplinary analysis, structural optimization, and design exploration (http://altairhyperworks.com/Default.aspx).
4. ANSA by BETA CAE Systems S.A. is a software that offers a complete solution for multidisciplinary CAE pre- and postprocessing. It is an advanced multidisciplinary CAE preprocessing tool that provides all the necessary functionality for full-model buildup, from Computer Aided Design (CAD) data to a ready-to-run solver input file, in a single integrated environment (https://beta-cae.com/ansa.htm).

5. ANSYS software by ANSYS, Inc. is a complex software allowing structural analysis that enables the user to solve structural engineering problems, thermal analysis, etc. With FEA tools, one can customize and automate simulations, and parameterize them to analyze multiple design scenarios. ANSYS Structural Mechanics software easily connects to other physics analysis tools, providing even greater realism in predicting the behavior and performance of complex products (http://ansys.com).
6. CFD and Multiphysics by ESI GROUP—Among the many options offered by the developer, it is one of the most interesting as it simulates the interaction of liquids and gases with surfaces by performing millions of numerical calculations. CFD analysis is mainly carried out earlier in the design process, based on which the prototype is made. With high-speed supercomputers, better solutions can be achieved. Multiphysics is basically advanced CFD involving multiple or coupled physics; for example, combining chemical kinetics and fluid mechanics or combining finite elements with molecular dynamics. Multiphysics typically involves solving coupled systems of partial differential equations. CFD-FASTRAN is the leading commercial CFD software for aerodynamic and aerothermodynamic applications. It is specifically designed to support the aerospace industry. It employs a state-of-the-art multiple moving body capability for simulating the most complex aerospace problems (http://esi-group.com/software-services/).
7. COMSOL Multiphysics FEA Software by COMSOL, Inc. is a software allowing electrical, mechanical, fluid flow, chemical modeling, and simulation (https://comsol.com/comsol-multiphysics).
8. Computer Simulation Technology (CST) STUDIO SUITE is an electromagnetic (EM) simulation software comprising several modules, namely MICROWAVE STUDIO®, for the fast and accurate 3D simulation of high-frequency devices (antennas, filters, couplers, planar, and multilayer structures, etc.); CST EM STUDIO® is used for the design and analysis of static and low-frequency EM applications such as motors, sensors, actuators, transformers, and shielding enclosures; and CST PARTICLE STUDIO® has been developed for the fully consistent simulation of free-moving charged particles. Applications include electron guns, cathode ray tubes, magnetrons, and wake fields; CST CABLE STUDIO® is for the simulation of signal integrity and electromagnetic compatibility (EMC) and electromagnetic interference (EMI) analysis of cable harnesses; CST printed circuit board (PCB) STUDIO® is for the simulation of signal integrity and EMC/EMI EMI on PCBs; CST MPHYSICS® STUDIO is for thermal and mechanical stress analysis; and CST DESIGN STUDIO™ is

a versatile tool that facilitates 3D EM/circuit cosimulation and synthesis (https://cst.com/Products).

9. FEMGV by TNO DIANA BV is a general-purpose pre- and postprocessor for FEA and CFD analysis software. DIANA, standing for Displacement Analyzer, is the main product of the company, which is an extensive multipurpose finite element software package that is dedicated to a wide range of engineering problems (http://tnodiana.com/).
10. Diffpack, provided by inuTech GmbH, is an object-oriented problem-solving environment for the numerical solution of PDEs. It can be utilized when dealing with the problem of stationary and time-dependent PDEs, linear and nonlinear PDEs, systems of coupled PDEs (multiphysics problems), finite element algorithms, mixed finite element algorithms, and implicit and explicit finite difference (www.inutech.de).
11. Femap by Siemens PLM Software is an advanced engineering simulation software that is able to create FEA models of complex engineering products and systems, and displays solution results. Femap can virtually model components, assemblies, or systems and determine the behavioral response for a given operating environment, including linear analysis with applied loads and constraints that are static, nonlinear static, and dynamic: effects due to contact (where one part of the model comes into contact with another), nonlinear material definitions (plasticity, elasticity, etc.), and large displacement (strains that exceed small displacement theory that limits a linear analysis approach); normal modes: natural frequencies of vibration; dynamic response: loads or motions that vary with time and frequency; buckling: critical loads at which a structure becomes unstable; and heat transfer: conduction, radiation, and phase change (http://plm.automation.siemens.com/en_us/plm/index.shtml).
12. FEMtools by Dynamic Design Solutions NV is a tool for static and dynamic simulation, verification, validation, and updating of finite element models. It also includes modules for structural optimization and for obtaining experimental reference data. It is not meant for textiles but rather for aerospace—frames, wings, reaction engines, automotive; biomechanics—orthopedic implants, prosthetic devices; civil engineering—bridges, dams; electronics/electrical—fans, cover plates, chassis, vibration; medical technology—bone material identification, dynamic problems in bones and prosthetics (http://femtools.com/products/index.htm).
13. FlexPDE by PDE Solutions, Inc. is an original scripted multiphysics finite element solution environment for PDEs that can model heat

flow, perform stress analysis, and model diffusion, electromagnetic field, etc. (http://pdesolutions.com/).

14. HyperFEA by Collier Research Corporation is a design software specializing in composite materials; integrating with other software, it uses FEA computed element forces to optimize a structure's cross-sectional dimensions and select the optimum materials and composite layups. The software is meant for aerospace, space launch, wind energy, high-speed rail, or ship-building (http://hypersizer.com/).
15. Integrated Engineering Software, Inc. (IES) offers VisualAnalysis, apart from Shape Builder and many other solutions. It is an easy-to-use tool for structural analysis and design for frames, trusses, tanks, and foundations (http://iesweb.com/).
16. IMPETUS AFEA Solver by IMPETUS Afea AS is a system for nonlinear explicit finite element simulations. It is primarily developed to predict large deformations of structures and components exposed to extreme loading conditions (http://impetus-afea.com/).
17. TSV-Solutions by TechnoStar Co., Ltd. is a set of applications meant for FEA of different engineering problems (e.g., acoustic FEM analysis, or an automobile engine cylinder block model) (http://en.e-technostar.com/products/).
18. JMAG by JSOL Corporation is a high-precision FEA software meant for various engineering problems but mainly for magnetic field analysis, heat induction, wireless power transfer, etc. (https://jmag-international.com/products/index.html).
19. LS-DYNA® by Livermore Software Technology Corporation is a general-purpose finite element program capable of simulating complex real-world problems. It is used by the automobile, aerospace, metal forming, construction, military, manufacturing, and bioengineering industries, specifically in electromagnetics, CFD, metal forming, etc. (http://lstc.com/products/ls-dyna).
20. Actran, Adams, Adams Machinery, Digimat Easy, Marc, Simufact Dytran, the MacNeal-Schwendler Corporation (MSC) Fatigue, MSC Nastran, Sinda, and others by MSC Software are powerful, different-purpose, nonlinear FEA solutions to accurately simulate the product behavior under static, dynamic, and multiphysics loading scenarios. For example, Marc's versatility in modeling nonlinear material behaviors and transient environmental conditions makes it ideal to solve complex design problems. Marc can be successfully utilized in acoustics, composites, crash and safety simulations, design optimizations, fatigue and durability, multibody dynamics, noise and vibrations propagation, structural analysis and thermal analysis, etc. MSC Nastran is a multidisciplinary structural analysis application

used by engineers to perform static, dynamic, and thermal analysis across the linear and nonlinear domains, complemented with automated structural optimization and award-winning embedded fatigue analysis technologies, all enabled by high-performance computing (http://mscsoftware.com/).

21. TRUE multiscale™ technology by MultiMechanics is a multipurpose software allowing modeling and analysis in the field of materials science and continuum mechanics. It focuses on building cutting-edge tools for the analysis of advanced and heterogeneous materials. Its advantage lies in its TRUE multiscale technology, which extends the flexibility and robustness of FEA to the microstructural level. This software is capable of generating complex 2D and 3D woven microstructures, braided structures, composites, etc. This software is really meant for textile engineers (http://multimechanics.com/).
22. NEi Nastran, NEi Editor, NEi Composites, NEi Fusion, NEi Works Basics and Works Expert, NEi Explicit, etc. by NEi Software are highly accurate, industry-proven solvers for solution generation. Among all of these, NEi Software can provide solutions for composite analysis solutions that allow product and design engineers to examine the structural, dynamic, and thermal aspects of their design with detailed graphical visualization of data and results (http://nenastran.com/).
23. S-Frame Analysis, S-View, S-Steel Design, S-Pad Design, S-Foundation Design, S-Concrete Design, S-Line Design, and S-Calc by S-Frame Software, Inc. are a set of sophisticated software for civil and structural engineers. The main applications are braced frames, trusses, bridges, office and residential buildings, skyscrapers, but also fabric structures, cable structures and more (https://s-frame.com/index_files/ProductList.htm).
24. SolidWorks by Dassault Systèmes SolidWorks Corporation is a set of tools offering a complete 3D environment that allows the creation, simulation, and management of different types of models.

    It is a multipurpose tool, useful in testing the modeled objects against a broad range of parameters during the design process, such as durability, static and dynamic response, assembly motion, heat transfer, fluid dynamics, and plastics injection molding. It is fully compatible with Abaqus CAE. It can be useful in textiles modeling (http://solidworks.com/).
25. WiseTex by the Composite Material Group of Katholieke Universiteit Leuven in collaboration with other universities and industrial partners, develops the software for modeling textiles and textile composites. It is meant for the creation of the internal geometry of dry and impregnated textiles (modules: WiseTex, LamTex, WeftKnit),

virtual reality visualization of the textile geometry (VRTex), resistance of textiles to tension, shear, and compaction (WiseTex), permeability of textiles (FlowTex), and stiffness of textile composites (TexComp) (https://mtm.kuleuven.be/Onderzoek/Composites/software/wisetex).

26. NISA by Cranes Software, Inc. is a comprehensive engineering analysis suite available to address the automotive, aerospace, energy and power, civil, electronics, and sporting goods industries. NISA can be used in estimating temperature distribution, heat flux, heat flow, etc. in steady and transient conditions under the influence of heat transfer by conduction/convection/radiation and in analyzing 3D incompressible and 3D compressible fluid flow, such as steady/transient, laminar/turbulent, Newtonian/non-Newtonian, subsonic/transonic/hypersonic, inviscid/viscous, and chemically reacting flows with mixing, as well as in the analysis of strength and failure behavior of composite materials under a variety of loads and environmental conditions (http://nisasoftware.com/services/mechanical-services).

List containing the aforementioned software packages and others are available online (List of finite element software packages, 2016).

References

Getting Started with Abaqus: Interactive Edition for Abaqus 6.13, 2016 by Dassault Systèmes, available online.

List of finite element software packages available at http://globalstressengineers.info/2013/02/list-of-finite-element-software-packages/.

2

Basics of Abaqus CAE Software

2.1 Creation of Models—Step-by-Step

As soon as Abaqus Computer-Aided Engineering (CAE) software also known as Complete Abaqus Environment is launched, the first option called *New* should be automatically opened. If this is not the case, the user should select the icon New situated in the top menu bar. In both the mentioned cases, a comment "A new model database has been created" will appear in the message area (Figure 2.1).

From now on, the user should follow the *Model tree*, which is situated on the left side of the screen and is presented in Figure 2.1. The term *Model tree* is to be utilized in the whole book, which can be treated as a reference area allowing for step-by-step model creation, knowing what the stage of the modeling process is and being able to switch between different modules of the *Model tree* if needed.

The crucial stages allowing creation of the model are as follows:

1. Creation of a part
2. Materials
3. Section and its properties
4. Defining the assembly
5. Creation of a step
6. Data output request
7. Boundary conditions (BCs) and loads
8. Meshing and element types
9. Job.

These nine listed stages are probably saying nothing to the user beyond being elements of the puzzle at the moment. Each of them is presented in a separate subchapter in a concise manner below Figure 2.1. As you may notice, the *Model tree* contains many different stages, apart from these nine

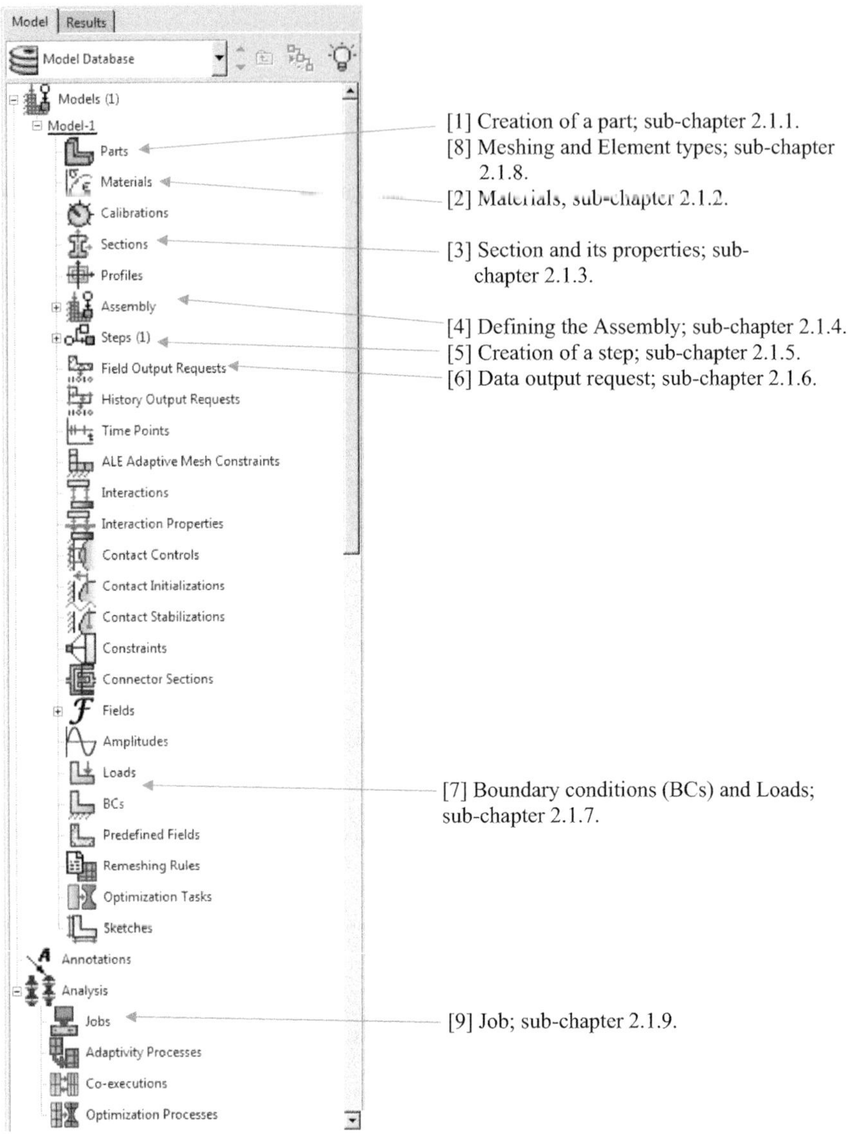

FIGURE 2.1
A *Model tree* from Abaqus CAE. © Dassault Systèmes, a French "société européenne" (Versailles Commercial Register # B 322 306 440), or its subsidiaries in the U.S. and/or other countries.

listed earlier. However, going through these nine stages allows the user to create a basic model. If necessary, one should elaborate other elements of the *Model tree* (e.g., *Interactions, Contact Controls*, and *Constraints*), although utilizing this option depends purely on the modeled objects or phenomena.

2.1.1 Creation of a Part

There are three options allowing the user to operate with the geometry of an object in Abaqus CAE. By *geometry*, one understands the shape of an object or a collection of objects with different elements that represent the shape of the thing that is to be modeled. The first option for having any geometry in the software is to import an already-prepared geometry, or a sketch, which is the basis for further modeling. In fact, one may import a *Sketch*, a *Part*, an *Assembly*, and a *Model*, so importing different types of files at different stages is possible. This usually takes place when the geometry is far too complex to try to create it on one's own using the graphical tools in Abaqus CAE or it has been prepared using a specific software. In many cases, it means importing a file prepared in other software (e.g., SolidWorks or computer-aided three-dimensional interactive application, CATIA). Sketches, Parts, and Assemblies or Models are characterized by certain features and have some limitations (e.g., a Sketch is literally a sketch, a drawing, or a shape composed of lines). In the case of importing a *Sketch*, one may select the following file types: ACIS SAT (.sat), IGES (.igs, .iges), STEP (.stp, .step), and AutoCAD DXF (.dxf). In the case of importing a *Part*, one may select the following file types: ACIS SAT (.sat), IGES (.igs, .iges), VDA (.vda), STEP (.stp, .step), CATIA V4 (.model, .catdata, .exp), CATIA V5 (.CATPart, .CATProduct), Parasolid (.x_t, .x_b, .xmt), ProE/NX Elysium Neutral (.enf), Output Database (.odb), and Substructure (.sim). In the case of importing an *Assembly*, one may select the following file types: Assembly Neutral (.eaf), CATIA V4 (.model, .catdata, .exp), Parasolid (.x_t, .x_b, .xmt), and ProE/NX/CATIA V5 Elysium Neutral (.enf). In the case of importing a *Model*, one may select the following file types: Abaqus/CAE Database (.cae), Abaqus Input File (.inp, .pes), Abaqus Output Database (.odb), Nastran Input File (.dbf, .dat, .nas, .nastran, .blk, .bulk), and Ansys Input File (.cdb).

The second option is to have the object scanned if its geometry is very complex because it is too difficult and/or too time consuming to sketch it in Abaqus CAE. The level of precision required from a model is also a meaningful factor when opting to have an object scanned first and next introduced into Abaqus as, for example, a stereolithography (STL) file format (*.stl). In many cases, it concerns elements of anatomy (e.g., hands, feet, and heart). Have your STL file ready, go to *Plug-ins/Tools/STL Import*.

The third option is simply to create your own geometry and ascribe to it all the required parameters to establish a model. To do so, one may select *Parts* in the *Model tree* or select the *Create Part* icon on the left side of the viewport, or by selecting *Part* and then *Create* from the top bar menu. As soon as one of these options is selected, a new submenu window appears.

This submenu, presented in Figure 2.2, is one of the key elements in the modeling process, as it allows the user to establish where the modeling process is to take place, in what space (*3D*, *2D Planar* or *Axisymmetric*), what type

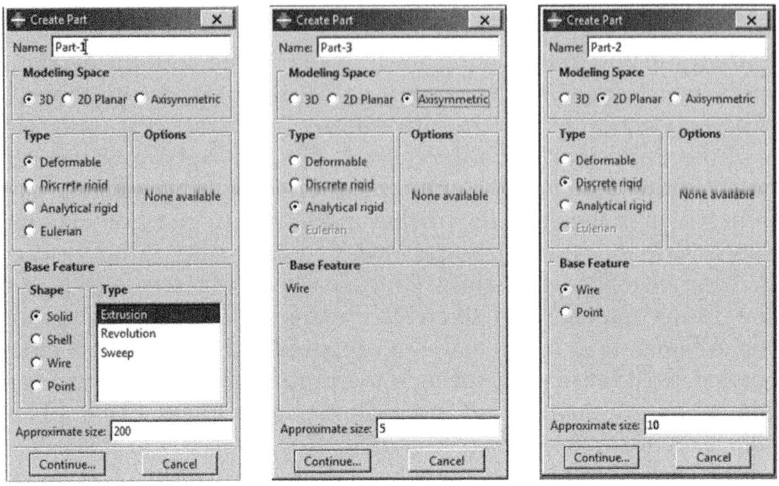

(a) (b) (c)

FIGURE 2.2
Examples of submenu *Create Part*, where specific options can be selected to establish a model (e.g., (a) three-dimensional deformable solid body type formed by extrusion, (b) two-dimensional discrete rigid model (rigid body can be made of nodes or elements), and (c) symmetrical about an axis analytical rigid model). © Dassault Systèmes, a French "société européenne" (Versailles Commercial Register # B 322 306 440), or its subsidiaries in the U.S. and/or other countries.

of body one wishes to deal with (*Deformable, Discrete rigid, Analytical rigid, Eulerian*), what form/shape of model one wishes to create (*Solid, Shell, Wire, Point*), how this model is to come into being, in what operation it is to be created (*Extrusion, Revolution, Sweep*), and what the size of the background grid is, being a canvas on which a drawing/sketch of the geometry of the model may take place (*Approximate size*). Depending on the requirements and needs, one may select 3D or 2D Planar or Axisymmetric, which changes the potential selection of other options, as presented in Figure 2.2a–c.

The following set of tools no. 1 can be used to create the appropriate geometry:

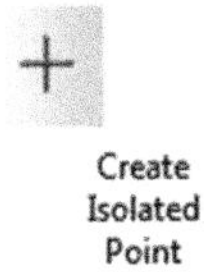

Icon representing *Create Isolated Point* tool. Each time one indicates any tool, the information about its function will appear.

Icon representing *Create Lines: Connected* tool

Icon representing *Create Circle: Center and Perimeter* tool

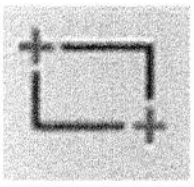
Icon representing *Create Lines: Rectangle (4 Lines)* tool

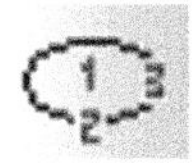
Icon representing *Create Ellipse: Center and Perimeter* tool

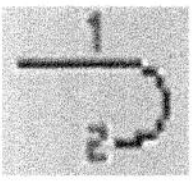
Icon representing *Create Arc: Tangent and Adjacent Curve* tool

Icon representing *Create Arc: Center and 2 Endpoints* tool

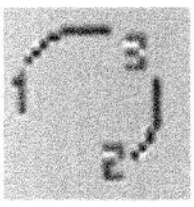
Icon representing *Create Arc: Thru 3 Points* tool

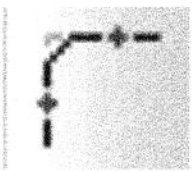
Icon representing *Create Fillet: Between 2 Curves* tool

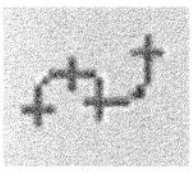
Icon representing *Create Spline: Thru Points* tool

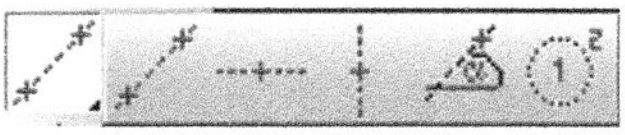
A group of icons representing, respectively, *Create Construction: Oblique Line Thru 2 Points, Horizontal Line Thru Point, Vertical Line Thru Point, Line at an Angle, Circle* tools

Two icons representing, respectively, *Set as Construction* and *Unset Construction* tools

Two icons representing, respectively, *Project Edges* and *Project References* tools

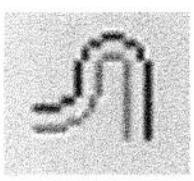
Icon representing *Offset Curves* tool

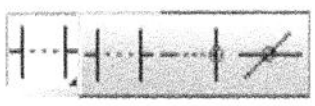
A group of icons representing, respectively, *Auto-Trim, Trim/Extend,* and *Split* tools

A group of icons representing, respectively, *Remove Gaps and Overlaps, Repair Short Edges,* and *Merge Vertices* tools

A group of icons representing, respectively, *Translate*, *Rotate*, *Scale*, and *Mirror* tools

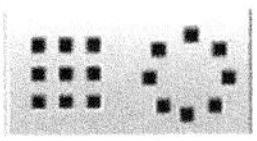

Two icons representing, respectively, *Linear Pattern* and *Radical Pattern* tools

Icon representing *Auto-Constrain* tool

Icon representing *Add Constraint* tool

Icon representing *Auto-Dimension* tool

Icon representing *Add Dimension* tool

Icon representing *Edit Dimension Value* tool

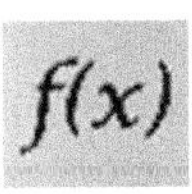

Icon representing *Parameter Manager* tool

Icon representing *Undo Last Action* tool

Icon representing *Redo Last Action* tool

Icon representing *Drag Entities* tool

Icon representing *Delete* tool

Icon representing *Add Sketch* tool

Icon representing *Save Sketch As* tool

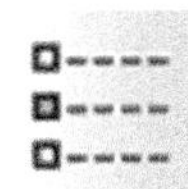

Icon representing *Sketch Options* tool. Selection of this tool opens new submenu Sketcher Options, where one may work with a *General* view on the sketch, its *Dimensions, Constraints,* and the *Image* itself.

Icon representing *Reset View* tool

Set of tools no. 1. Set of icons representing graphical tools. © Dassault Systèmes, a French "société européenne" (Versailles Commercial Register # B 322 306 440), or its subsidiaries in the U.S. and/or other countries.

The same tools can be reached using a toolbar at the top of the screen (marked as 2 in Figure 2.3) when the Abaqus CAE/*New* window has been opened and the model predefined in the *Create Part* submenu has been reached. However, it seems to be far more convenient to use icons instead of the tool bar at the top of the screen.

Let us select icon *Create Circle: Center and Perimeter* twice by sketching two circles—one smaller and one large—on the grid so that they become concentric circles (having a common center). After utilizing (e.g., the icon *Add Dimension*), one may dimension both circles. As soon as dimensions are ascribed to the circles, their shapes will be adjusted accordingly.

It is now time to mention how crucial the decision about the units of the introduced values is when dimensioning the geometry. The user needs to decide what set of units will be utilized consistently until the end of the model creation process. Abaqus CAE does not require units to be introduced into any submenu window that appears when modeling. It is the user who is responsible for introducing correctly converted values in order to receive the appropriate and correctly estimated results. The examples presented in this book utilize only the standard SI unit system. The reader can find out more about the units in the appropriate references (Getting Started with Abaqus: Interactive Edition for Abaqus 6.13, 2016 Dassault Systèmes).

At the same time, the following comment: "Sketch the section for the solid extrusion" and *Done* button will be displayed in the prompt area (marked as 7 in Figure 2.3). After selecting *Done*, another submenu appears, and the user can introduce the extrusion parameter, which is in this case a length of the tube. The consequences of sketching, dimensioning, and extruding are presented in Figure 2.3, where the most important regions of the screen that the user is working with are marked.

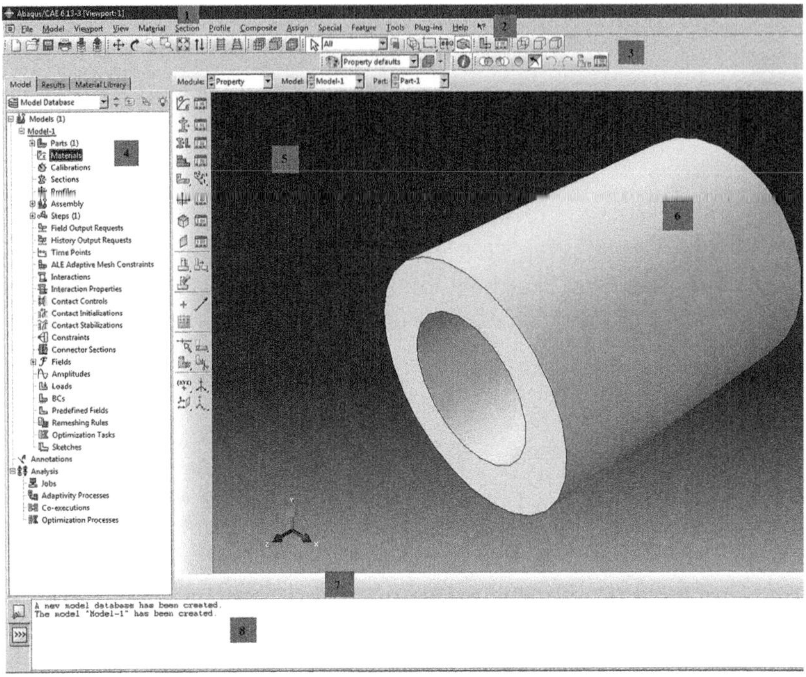

FIGURE 2.3
Main Abaqus/CAE window view with identified components, where [1] is the title bar, which presents the name and version of the software (Abaqus/CAE 6.13-3 [Viewport 1]). It usually presents the specific name of the file given by its author. In this case, the name of the file was not given; [2] is a menu bar, which contains all available tool options and functionalities of the software; [3] is a toolbar presenting small icons. These icons give access to some menu options (e.g., how to change the view on the cylinder) [6]. Just below [3] is a set of tabs consisting of Model, Results, and Material Library. The Model tab refers only to parameters of the model ascribed during the process of creation, while Results tab presents another set of options allowing review of the results and visualization. The Material Library tab allows the creation of different sets of materials; [4] is a *Model tree*, which presents the list of parameters/characteristics of the models starting with a graphical overview of the model and the objects that it contains, such as parts, materials, steps, loads, and output; [5] is a background field. The modeled object will always appear in this field. [6] is a modeled object. Figure 2.3 depicts the geometry of the cylinder. This is the initial moment of the modeling phase. The material parameters have not been ascribed to the geometry yet. [7] is a prompt area. [8] is a message area, where all the comments concerning the current status of processing of the model can be found (e.g., warnings or reasons why the processing could not be completed).

2.1.2 Materials

When we talk about *Materials* or material properties of the modeled objects, we are mainly thinking about the substance of which the model is made (e.g., what is the density of the material), its Young's modulus, Poisson's ratio, thermal conductivity, electrical conductivity, or sorption. For example, when modeling a woven fabric made of different warps and wefts that are additionally blends of different raw materials like cotton/polyester and polyamide/wool, the user may model warps and wefts—yarns, as homogeneous solid bodies and apply mechanical parameters for the model that reflect the properties of these yarns, rather than the specific raw materials. Editing the material properties can be done by selection of the *Materials* option in the *Model tree* on the left side of the screen, by selection of the appropriate icon, or by selecting *Material,* and next *Create* in the top toolbar.

A large group of materials, especially metals but also a large group of materials utilized for the production of textiles, may be treated as materials with approximately linear elastic characteristics at low strain magnitudes (Figure 2.4) with constant Young's or elastic modulus (E). Stress σ is defined here as the force acting on a selected area. Thus, the unit of σ is the same as the unit of pressure, namely Pascals [Pa]. A Pascal is a Newton [N] per square meter [m^2].

Thus, an elastic material can be characterized by a constitutive relation giving the stress as a function of current deformation. The mechanical characteristics of the elastic materials can be described by a linear stress (σ) and strain (ε) relation as presented in Figure 2.4.

If the applied stress is higher (stress and strain magnitude are high), these materials may start to behave like materials with nonlinear characteristics, thus representing the inelastic group of materials, which is usually called plastic material, or where the term plasticity of the material is utilized in relation with them (Figure 2.5).

In the case of other materials that cannot be treated as having linear elastic characteristics, as their stress–strain characteristics are nonlinear (although at low strain magnitudes they may present typically elastic behaviors), the term hyperelastic materials is utilized, where the character of the relation between σ and ε is not as simple as in the case of linear elastic materials, and in fact, different numerical approaches are utilized to define the nonlinear elastic response of the materials (e.g., rubber, gel, and human tissue, especially skin, arteries, and muscles). In fact, several different characteristics

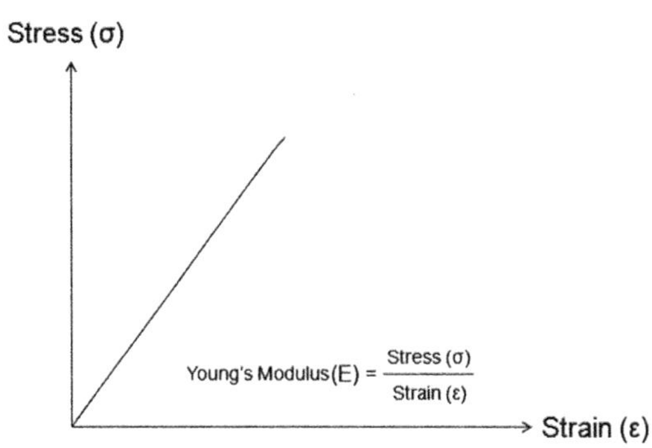

FIGURE 2.4
A typical linear stress–strain characteristic for elastic materials.

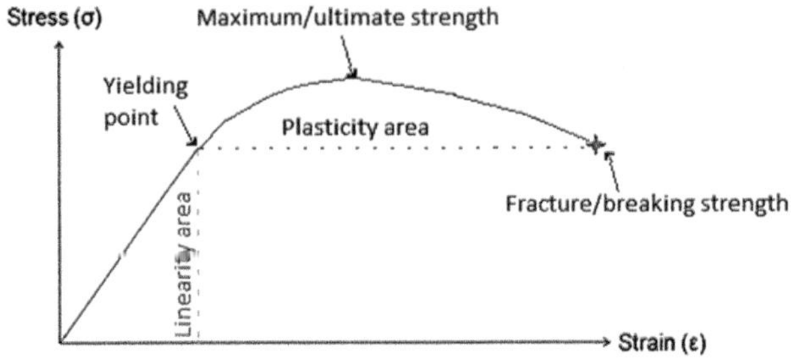

FIGURE 2.5
A typical stress–strain curve presenting both elastic and plastic characteristics of materials.

describing different hyperelastic materials have been elaborated for their nonlinear stress–strain relations, and a typical example is presented in Figure 2.6.

Instead of the typical stress–strain relation that is meant for elastic materials, hyperelastic materials are described in terms of their *strain–energy potential W*.

W defines the strain energy stored in the material per unit of reference volume of the material in its initial setup (configuration) as a function of the strain at that point in the material. There are several different formulas according to which this strain energy can be calculated, and the application of the specific formula depends on the type of hyperelastic material. The following formulas can be utilized in Abaqus CAE software in the case of isotropic hyperelastic materials, and their properties are characterized by the strain–energy (stored-energy) function:

a. Arruda–Boyce model (Arruda and Boyce, 1993).

   This model is based on the relation of an eight-chain representation of the underlying macromolecular network structure of rubber and the non-Gaussian behavior of the individual chains in the proposed network. The eight-chain model captures the relation in the network deformation while requiring only two material parameters: the initial modulus and a limiting chain extensibility.

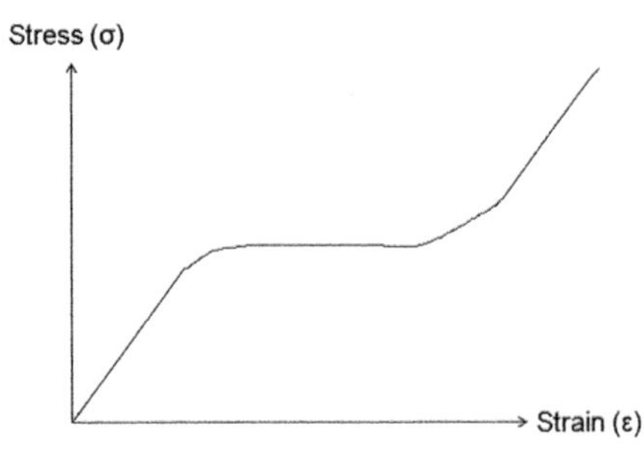

FIGURE 2.6
Typical, linear stress–strain characteristics for hyperelastic materials.

This model accurately represents rubber and other polymeric, rubber-like substances. This model presents the first five terms for the strain energy:

$$W = nk\theta\left(\frac{1}{2}(I_1 - 3) + \frac{1}{20N}\left(I_1^2 - 9\right) + \frac{11}{1050\ N^2}\left(I_1^3 - 27\right)\right.$$

$$\left. + nk\theta\left(\frac{19}{7000\ N^3}\left(I_1^4 - 81\right) + \frac{519}{673750\ N^4}\left(I_1^5 - 243\right)\right)\right) + \ldots$$

where W is the strain–energy potential, the strain energy presents a nonlinear I_1 dependence as a result of chain stretch, n is the chain density, k is Boltzmann's constant, θ is the temperature, I_1 is the first stretch invariant, and N is the number of rigid links composing a single chain. Due to the fact that each chain is subject to stretch, which is equivalent to that in every other network chain, the model utilizes average values for a single chain over eight spatial orientations. The first stretch invariant is $I_1 = \lambda_1^2 + \lambda_2^2 + \lambda_3^2$; the chain stretch can be reduced to the following expression, where only I_1 is utilized: $\lambda_{chain} = \frac{1}{\sqrt{3}} I_1^{\frac{1}{2}}$.

b. Marlow model (Getting Started with Abaqus: Interactive Edition for Abaqus 6.13, 2016 Dassault Systèmes; Marlow 2003; Marlow 2008)

This strain–energy potential model was implemented in Abaqus CAE (R.S. Marlow).

The model of the Marlow strain–energy potential proposed by Abaqus CAE is

$$U = U_{\text{dev}}\left(\bar{I}_1\right) + U_{\text{vol}}\left(J_{\text{el}}\right),$$

where U is a strain–energy potential (U is interchangeable with W in the literature) stored in the material per unit of reference volume of this material; U_{dev} represents a deviatoric component of a strain energy, and U_{vol} is a volumetric component of strain energy; and $\bar{I}_1$ is the first deviatoric strain invariant being a sum of deviatoric stretches (it is the equivalent of I_1 from Arruda–Boyce model).

Deviatoric stretches $\bar{\lambda}_i = J^{\frac{-1}{3}}\lambda_i$; where J is the total volume ratio, J_{el} is the elastic volume ratio, and λ_i is a principal stretch (λ_1, λ_2, λ_3 as in Arruda–Boyce model).

c. Mooney–Rivlin model (Ali et al., 2010, Gent, 2012; Getting Started with Abaqus: Interactive Edition for Abaqus 6.13, 2016 Dassault Systèmes; Ruíz and González, 2006; Rivlin, 1948).

Its disadvantage is the fact that material parameters must be obtained by carrying out experiments. Mooney–Rivlin model is meant for hyperelastic incompressible materials, where the strain–energy potential function W (sometimes also called as strain–energy density) is a linear combination of two invariants of the left Cauchy—Green deformation tensor. The model of Mooney–Rivlin is as follows:

$$W = C_1\ (I_1 - 3) + C_2\ (I_2 - 3)$$

where W is a strain–energy potential per unit of the reference volume; C_1 and C_2 are empirically determined material constants; and I_1 and I_2 are the first and second invariants of the unimodular component of the left Cauchy–Green deformation tensor, respectively.

d. Neo-Hookean model (Ali et al., 2010; Getting Started with Abaqus: Interactive Edition for Abaqus 6.13, 2016 Dassault Systèmes; Ruíz and González, 2006).

This model is meant for rubbers with limited compressibility.

$$W(I_1) = C_{10}(I_1 - 3)$$

where W is a strain–energy potential per unit of the reference volume, C_{10} is a material constant, and I_1 is the first deviatoric strain invariant.

The fully incompressible limit can be obtained by setting in the stress–strain law. The energy–strain function expression utilizes strain invariant I_1. This model has the simplest hyperelastic model for elastomeric materials when material data is insufficient.

e. Ogden model (Ali et al., 2010; Bergström, 2015; Getting Started with Abaqus: Interactive Edition for Abaqus 6.13, 2016 Dassault Systèmes; Ogden, 1972).

The model is meant for solid rubber-like incompressible materials (e.g., rubber, biological tissues, and selected polymers), which possess a strain–energy function and which are isotropic relative to the stress-free ground state. The principal stretches are connected via $\lambda_1\lambda_2\lambda_3 = 1$. There are different ways of writing the model—in case of compressible representation of the material, one should utilize the following equation for W:

$$W(\lambda_1, \lambda_2, \lambda_3) = \sum_{k=1}^{N} \frac{2\ \mu_k}{\alpha_k^2}\left((\lambda_1)^{\alpha_k} + (\lambda_2)^{\alpha_k} + (\lambda_3)^{\alpha_k} - 3\ \right)$$

where W is a strain–energy potential per unit of the reference volume, and N, μ, and α are material constants, which can be determined by experimental test.

f. Polynominal model (Ali et al., 2010; Getting Started with Abaqus: Interactive Edition for Abaqus 6.13, 2016 Dassault Systèmes).

This model of strain–energy function is usually used in modeling of elastomers.

$$W = \sum_{i+j=1}^{N} C_{ij}(\bar{I}_1 - 3)^i (\bar{I}_2 - 3)^j + \sum_{i=1}^{N} \frac{1}{D_i} (J-1)^{2i}$$

where W is a strain–energy potential per unit of the reference volume; $C_{i,j}$ and D_i are material coefficients in the polynomial; $\bar{I}_1$ and $\bar{I}_2$ are the first and second invariants of the left isochoric Cauchy–Green deformation tensor, respectively; and J is the elastic Jacobian (elastic volume ratio).

This model is also called the generalized Rivlin model.

$$\bar{I}_1 = \bar{\lambda}_1^2 + \bar{\lambda}_2^2 + \bar{\lambda}_3^2 \text{ and } \bar{I}_2 = \bar{\lambda}_1^{-2} + \bar{\lambda}_2^{-2} + \bar{\lambda}_3^{-2}$$

g. Reduced polynominal model (Getting Started with Abaqus: Interactive Edition for Abaqus 6.13, 2016 Dassault Systèmes);

This model is based on the first invariant.

$$W = \sum_{i+j=1}^{N} C_{ij} (\bar{I}_1 - 3)^i + \sum_{i=1}^{N} \frac{1}{D_i} (J-1)^{2i}$$

where W is a strain–energy potential per unit of the reference volume; $C_{i,j}$ and D_i are material coefficients in the polynomial; $\bar{I}_1$ is the first invariant of the left isochoric Cauchy–Green deformation tensor; and J is the elastic Jacobian (elastic volume ratio).

h. Van der Waals' model (Ali et al., 2010; Getting Started with Abaqus: Interactive Edition for Abaqus 6.13, 2016 Dassault Systèmes;)

$$W = \mu \left\{ -\left(\lambda_m^2 - 3\right)\left[\ln(1-\eta) + \eta\right] - \frac{2}{3} a \left(\frac{\ddot{I} - 3}{\lambda_m^2 - 3} \right)^{\frac{3}{2}} \right\} + \frac{1}{D_i} \left(\frac{J^2 - 1}{2} - \ln J \right)$$

where W is a strain–energy potential per unit of the reference volume; μ is the initial shear modulus; λ_m is the locking stretch, which is the stretch at which the stress goes to infinity because the polymer network chains are fully extended and rigid; $\eta = \sqrt{\frac{\ddot{I} - 3}{\lambda_m^2 - 3}}$ and $\ddot{I} = (1 - \beta)\bar{I}_1 + \beta \bar{I}_2$, where η is a parameter connecting deviatoric strain invariants with a locking stretch parameter; β is an invariant mixture parameter existing in the formula where another invariant parameter, $\ddot{I}$, combines the first and second deviatoric strain invariants; and D_i is a material compressibility parameter.

i. Yeoh model (Ali et al., 2010; Getting Started with Abaqus: Interactive Edition for Abaqus 6.13, 2016 Dassault Systèmes; Yeoh, 1993)

The effect of the second-order invariance on the general polynomial series expansion was studied, and it was revealed that the sensitivity of the strain–energy potential function W to changes in the second invariant is generally much smaller than the sensitivity to changes in the first invariant. Thus, it was omitted when building the strain–energy potential function equation. The Yeoh model of W can be considered as a special case of a reduced polynomial with a value of $N=3$.

$$W = \sum_{i+j=1}^{3} C_{i0} \left(\bar{I}_1 - 3\right)^i + \sum_{i=1}^{3} \frac{1}{D_i} \left(J - 1\right)^{2i}$$

where W is a strain–energy potential per unit of the reference volume; $C_{i,j}$ and D_i are temperature-dependent material parameters; $\bar{I}_1$ is the first invariant of the left isochoric Cauchy–Green deformation tensor; and J is the elastic Jacobian (elastic volume ratio).

The following formulas can be utilized by Abaqus CAE software in case of anisotropic hyperelastic materials, and their properties are characterized by strain–energy function:

a. Fung-Anisotropic model (Fung 1993; Getting Started with Abaqus: Interactive Edition for Abaqus 6.13, 2016 Dassault Systèmes; Sacks, 1999).

The Fung-anisotropic hyperelastic model is commonly used to model both engineered and native soft tissues used in medical device and surgical applications (e.g., human skin and veins).

The generalized Fung strain–energy potential has the following form:

$$W = \frac{c}{2}\left[e^{Q} - 1\right] + \frac{1}{D_i}\left[\frac{J^2 - 1}{2} - \ln J\right]$$

where W is a strain–energy potential per unit of the reference volume; $Q = A_1E_{11}^2 + A_2E_{22}^2 + 2A_3E_{11}E_{22} + A_4E_{12}^2 + 2A_5E_{12}E_{11} + 2A_6E_{12}E_{22}$, c and A_i are material constants. Equation defining Q is a generalized pseudoelastic Fung constitutive models, E_{ij} is Green's strain tensor in the material axes coordinate system, D_i is a material compressibility parameter, and J is the elastic Jacobian (elastic volume ratio).

b. Fung-Orthotropic model (Fung et al., 1979; Getting Started with Abaqus: Interactive Edition for Abaqus 6.13, 2016 Dassault Systèmes; Humphrey, 1995)

The strain–energy function was proposed by Fung et al. (1979) and developed in a more general form by Humphrey (1995). The strain—energy potential W can be expressed as

$$W(E) = W_{\text{Fung}} + U(J)$$

where W_{Fung} equals $c\left[e^{Q} - 1\right]$ and $U(J)$ is the volumetric energy utilized in finite element models computation being expressed as

$$U(J) = \frac{1}{2}\kappa(J-1)^2 \text{ and } \kappa = \frac{C}{3(1-2\upsilon)}$$

where J is the elastic Jacobian (elastic volume ratio), κ is a bulk modulus, c is a material constant having a dimension of a modulus, υ is Poisson's ratio, and

$$Q = A_1\ E_{11}^2 + A_2E_{22}^2 + A_3E_{33}^2 + 2\ A_{12}E_{11}E_{22} + 2\,A_{23}E_{22}E_{33}$$

$$+ 2\,A_{13}E_{11}E_{33} + 4\,A_{44}E_{12}^2 + 4\,A_{55}E_{23}^2\ + 4\,A_{66}E_{31}^2$$

c. Holzapfel Model (Bergström, 2015; Gasser et al., 2006; Getting Started with Abaqus: Interactive Edition for Abaqus 6.13, 2016 Dassault Systèmes; Holzapfel et al., 2000)

   The full name of this model that the reader may encounter in the literature is Holzapfel–Gasser–Ogden model, which is a strain–energy potential model for anisotropic hyperelastic materials, mainly for human tissues (e.g., veins). Like other listed models, it is also available in Abaqus software. This model is a combination of a Neo-Hookean model characterized earlier in this chapter and up to three groups of collagen fibers distributed in the veins. Different spatial orientation of the fibers can be presented as vectors:

$$\left[a_{1x}; a_{1y}; a_{1z}\right], \left[a_{2x}; a_{2y}; a_{2z}\right], \left[a_{3x}; a_{3y}; a_{3z}\right].$$

   Arteries are highly deformable composite structures which demonstrate a nonlinear stress–strain response with an exponential stiffening effect at higher pressures. This stiffening effect, common to all biological tissues, is based on the recruitment of embedded, load-carrying collagen fibers, which lead to the anisotropic mechanical behavior of arteries.

The strain–energy potential function can be presented as

$$W = \frac{\mu}{2}\left(\bar{I}_1 - 3\right) + \frac{k_1}{2k_2}\sum_{i=1}^{3}[e^{k_2\left(E_i^2\right)} - 1] + \frac{\kappa}{2}\,(J-1)^2$$

where W is a strain–energy potential per unit of the reference volume; μ, k_1, k_2 and κ are material parameters and contribution of collagen to the overall response; $\bar{I}_1$ is the first deviatoric strain invariant; E_i is an energy term; and J is the elastic Jacobian (elastic volume ratio).

The energy term can be presented as

$E_i = d(\bar{I}_1 - 3) + (1 - 3d)\left[\bar{I}_{4i}^{*} - 1\right]$, where d is a parameter that characterizes the level of dispersion of energy in the fiber directions and $\bar{I}_{4i}^{*}$ is related to pseudoinvariants $\bar{C}$: A, where $\bar{C}$ is a measure of deformation of a body and A represents second-order tensors which characterize the vein wall structure.

To ascribe material properties to the created geometry, one selects the Materials icon from the toolset on the left side of the viewport or selects *Materials* from the *Model tree*, or selects *Material*, and next *Create* from the top toolbar menu. As soon as one of these options is selected, a submenu window called *Edit Material* appears. This is the right time to name the material and introduce its main parameters, which depend on the type of model that is to be created. For example, if the user wants to perform an analysis related to heat transfer, he/she needs to prepare a set of appropriate parameters (e.g., conductivity, heat generation, and specific heat), and in the case of electrical phenomenon modeling, one will be required to introduce electrical conductivity for the material/materials for which the model is composed. As mentioned earlier, values need to be introduced, bearing in mind the consistency and appropriate conversion of the units.

2.1.3 Section and Its Properties

In many cases, a whole modeled object is composed of different elements made of different materials. Each of these elements—sections of the model—needs to have specific materials assigned.

In order to create a so-called Section, the user needs to select the *Section* icon from the toolset on the left side of the viewport or select *Sections* from the *Model tree*, or select *Section*, and next *Create* from the top toolbar menu. As soon as one of these options is selected, a submenu window called *Create Section* appears. Again, depending on the kind of model the user is attempting to elaborate and work with, different selections in this window open different possibilities, as presented in Figure 2.7.

As soon as a selection is made, another submenu window appears, as presented in Figure 2.8, and the user is asked to assign a specific material to the section of the part. *Material-1* is a set of information about the specific material of the modeled object(s) that the user created in the previous step.

Select OK to finalize this step.

In order to complete the assignment of the selected section to the Material and Part, one needs to go back to *Parts* in the *Model tree* and select *Section Assignments*.

This evokes a command presented in the prompt area below the viewport with a model. The command is *Select a region to be assigned a section*.

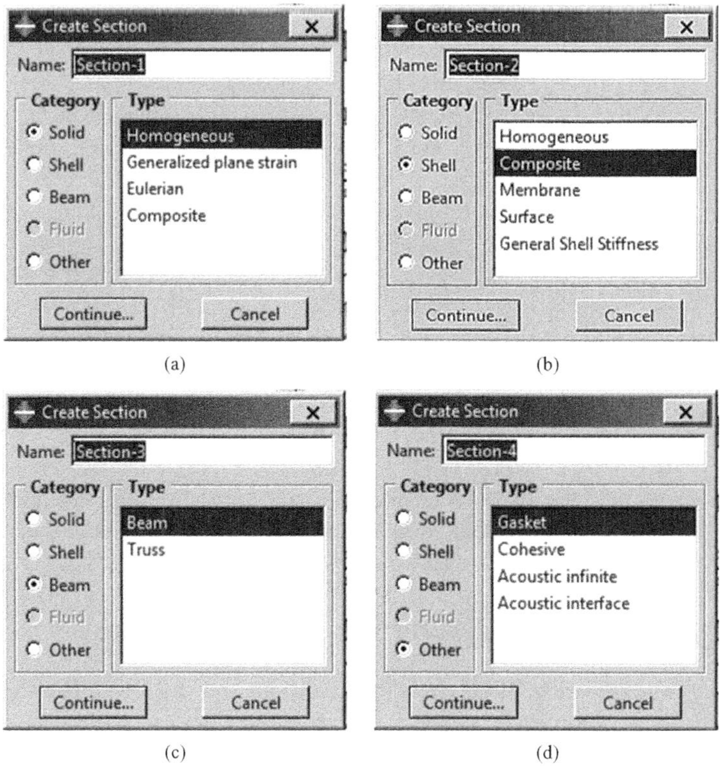

FIGURE 2.7
An exemplary selection of possible options in the submenu *Create Section*. © Dassault Systèmes, a French "société européenne" (Versailles Commercial Register # B 322 306 440), or its subsidiaries in the U.S. and/or other countries.

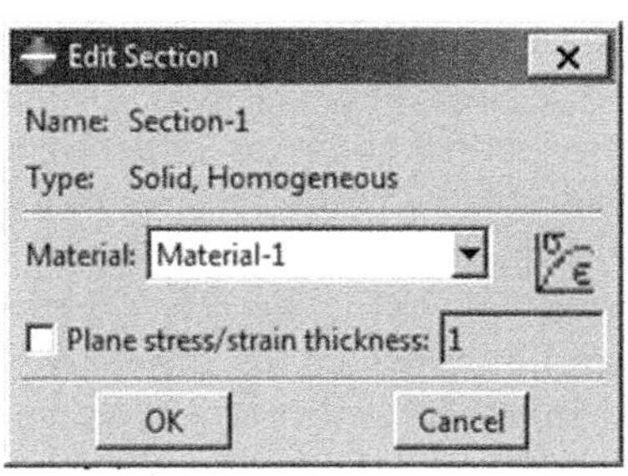

FIGURE 2.8
An exemplary submenu allowing assignment of the material (*Material-1*) to the section of the *Part* created in the previous step. © Dassault Systèmes, a French "société européenne" (Versailles Commercial Register # B 322 306 440), or its subsidiaries in the U.S. and/or other countries.

After selecting an appropriate region of the modeled object, which is immediately highlighted by Abaqus CAE, the user should select *Done*, which confirms the selection. As a consequence, a new submenu window *Edit Section Assignment* appears as presented in Figure 2.9, where the user needs to confirm the selection of the appropriate section and then select *OK*, which finalizes this step.

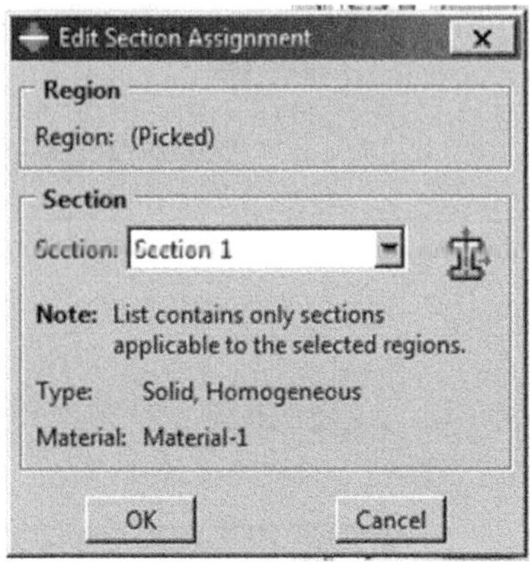

FIGURE 2.9
An exemplary submenu window *Edit Section Assignment* where *Section-1* was assigned. © Dassault Systèmes, a French "société européenne" (Versailles Commercial Register # B 322 306 440), or its subsidiaries in the U.S. and/or other countries.

After pressing OK, the object changes the color (in this specific case, it changes the color to light green; however, the selection of colors can be customized). The change of color is also a confirmation that a specific step was taken.

2.1.4 Defining the Assembly

Assembly is a collection of all *Parts* of the model, and all these *Parts* are present in the same coordinate system, which is not the case before defining an *Assembly*. Each of the Parts in the model exists in its own coordinate system and is independent from other parts in the model. Thus, a common platform, where all the parts may exist together having the same coordinate system, is needed. The creation of this platform containing different parts is called Defining the Assembly. In fact, the user defines a common platform for parts of the model by creating so-called Instances of a Part of the model and then placing these Instances in a common platform, such that they can coexist together, creating the required model in a single—global—coordinate system. Instances can be classified as either independent or dependent. Independent part instances are meshed individually, while the mesh of a dependent part instance is associated with the mesh of the original part.

In order to define the assembly, the user should go to the *Model tree* on the left side of the screen and select *Assembly* and double-click on *Instances* or select the *Create Instance* icon from the toolset on the left side of the viewport or select *Instance*, and next *Create* from the top toolbar menu. As soon as one of these options is selected, a submenu window called *Create Instance* appears, as presented in Figure 2.10. The user should select all the parts listed in this window to place them all in one coordinate system.

An option Auto-offset from other instances allows the user to double the number of Parts intended for instance. Independently, the user can multiply the number of instances and interspace them by using the following set of tools no. 2 presented on the left side of the screen:

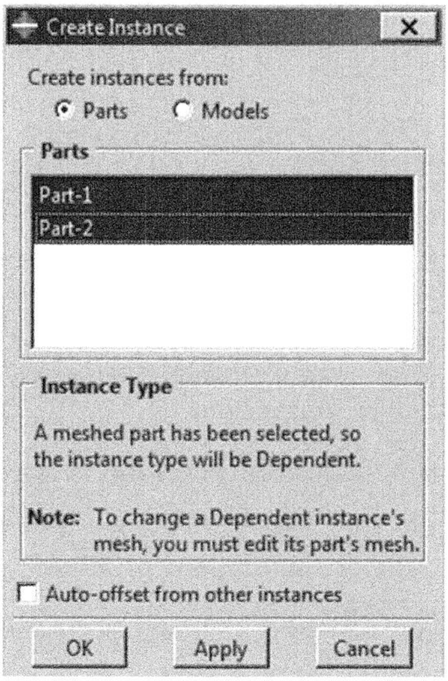

FIGURE 2.10
An exemplary submenu window *Create Instance*, where two parts are selected for an *Assembly*. © Dassault Systèmes, a French "société européenne" (Versailles Commercial Register # B 322 306 440), or its subsidiaries in the U.S. and/or other countries.

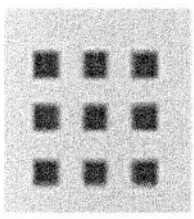
Icon representing *Linear Pattern* tool

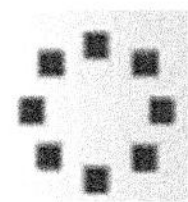
Icon representing *Radial Pattern* tool

Icon representing *Translate Instance* tool

Icon representing *Rotate Instance* tool

Icon representing *Translate To* tool

A group of icons representing, respectively, *Create Constraint: Parallel Face*, *Create Constraint: Face*

to Face, Create Constraint: Parallel Edge, Create Constraint: Edge to Edge, Create Constraint: Coaxial, Create Constraint: Coincident Point, and *Create Constraint: Parallel Csys* tools

Icon representing *Merge/Cut Instances* tool

Set of tools no. 2. Set of icons representing tools allowing operations on instances. © Dassault Systèmes, a French "société européenne" (Versailles Commercial Register # B 322 306 440), or its subsidiaries in the U.S. and/or other countries.

Press OK to confirm the selection. Both Parts appear in the same coordinate system.

2.1.5 Creation of a Step

A single *Step* in the model can be treated as a single step of an analytical protocol, a single moment of the scenario according to which the model is to perform.

By default, as soon as a *New* window for the modeling process appears, the first, *Initial* predefined step is already present. It can also be found by double-clicking on the *Steps (1)* option in the *Model tree,* as depicted in Figure 2.11.

An *Initial* step is a step created automatically, in which one applies interactions, BCs and predefined fields.

The analysis step—created by the user—depending on the needs and requirements, is a step in which one applies conditions and scenario steps, which are supposed to happen to the model (e.g., compressing).

There are two major types of steps in Abaqus CAE, as presented in Figure 2.12: *General,* which can be used to analyze the linear or nonlinear model response (e.g., coupled temperature–displacement, coupled thermal–electric), and *Linear perturbation,* which can be utilized exclusively to analyze linear problems (e.g., buckle, frequency, static linear perturbation, steady-state dynamics).

After selection, the user clicks on *Continue* and another submenu window appears. There are three tabs in this window: *Basic, Incrementation,* and *Other,* and different parameters are already predefined there. The user may introduce changes and define, for example, the type of incrementation: *Automatic*

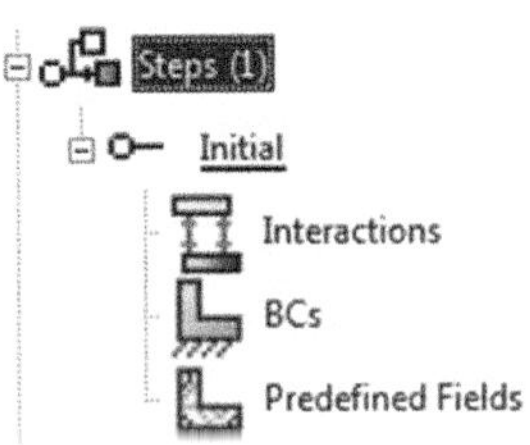

FIGURE 2.11
A fragment of the *Model tree* with a focus on the Initial step in the modeling process. © Dassault Systèmes, a French "société européenne" (Versailles Commercial Register # B 322 306 440), or its subsidiaries in the U.S. and/or other countries.

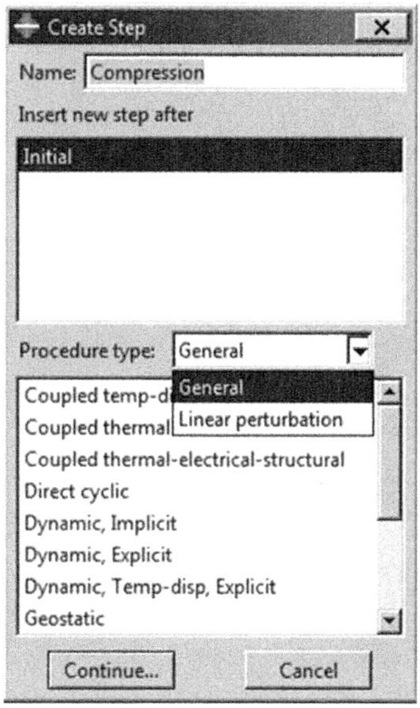

FIGURE 2.12
An exemplary submenu window *Create Step*, where a new step was named *Compression* and the type of the procedure *General* is selected. © Dassault Systèmes, a French "société européenne" (Versailles Commercial Register # B 322 306 440), or its subsidiaries in the U.S. and/or other countries.

or *Fixed*, the number of incrementations in which the analysis will take place, and the solution will be provided.

Press OK to confirm the selection.

2.1.6 Data Output Request

In this stage of the modeling process, the user indicates the set of parameters that will be delivered by the software after processing the model. For example, if the user models an object subject to strain under the influence of a load, he/she is interested in both the overall stress and strain parameters, but perhaps also in the specific stress and strain and probably in the displacement that occurs after applying a load, or maybe in the reaction forces that occur. Thus, there is the opportunity to select all the parameters that may be useful from the point of view of interpretation and discussion of the final results of modeling.

After finalizing *Step* creation, the *Field Output Request* brings up a default file that is a composition of the different parameters that are normally calculated for a specific Step, as presented in Figure 2.13.

To examine what the default selection is, the user is required to select the *Field Output Request* in the *Model tree* by using the right mouse button, and next by selecting *Manager*. A new submenu window called *Field Output Requests Manager*

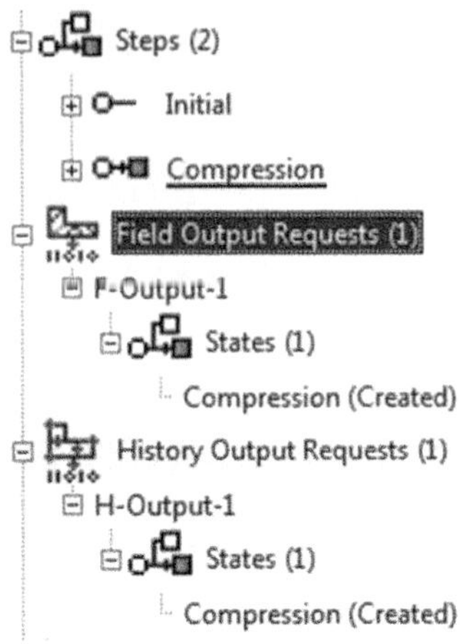

FIGURE 2.13
A fragment of the *Model tree* with a focus on the Field Output Request in the modeling process.

appears, and after selection of *Edit* in this new window, the user may observe a default selection in the *Edit Field Output Request* submenu window, as presented in Figure 2.14, and change it according to his or her own requirements.

In order to define *Field Output Requests,* one should go to the *Model tree* on the left side of the screen and select *Field Output Request* or select *Output,* and then *Field Output Requests* and *Create* from the top toolbar menu or to select the icon representing Create Field Output. As soon as one of these options is selected, a submenu window called *Create Field* appears, and after confirmation of a selection, a new submenu window called *Edit Field Output Request,* similar to the one presented in Figure 2.14, appears. In the case of creating a new *Field Output Request,* the automatically generated name should be F-Output-2 and no variables are selected.

In the case of managing the existing file, starting from the top of Figure 2.14, the user can find the default name of the file *F-Output-1,* the name of the step Compression, created previously. If there are other steps, they are listed here.

Next, the *Domain* menu is shown, which allows selection of the specific region of the model (e.g., composite layer) for which the output is to be generated. In the case of *Frequency,* the user can select which incrementation the output is to be presented for or which moment of the modeling period time the output is to be presented for. For example, selection of *Every n increment* does not activate the *Timing* dialog, but this box is activated after the selection of *Every x units of time* in the *Frequency* dialog box.

Next, Output Variables selection opportunities and a list of abbreviations of output variables presented in alphabetical order can be seen. These are as follows: CDISP, which stands for contact displacements; CF—concentrated forces and moments, CSTRESS—contact stresses; LE—logarithmic strain

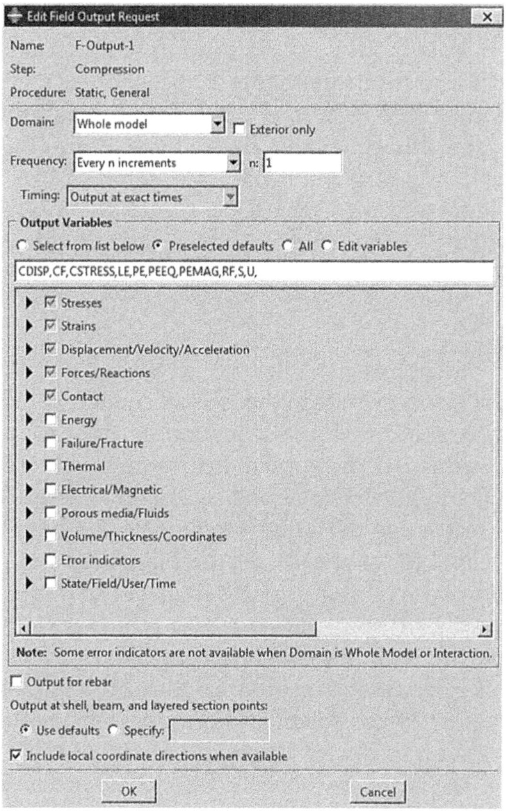

FIGURE 2.14
An exemplary submenu window *Edit Field Output Request.* © Dassault Systèmes, a French "société européenne" (Versailles Commercial Register # B 322 306 440), or its subsidiaries in the U.S. and/or other countries.

components; PE—plastic strain components; PEEQ—equivalent plastic strain; PEMAG—plastic strain magnitude; RF—reaction forces and moments; S—stress components and invariants; and U—translations and rotations. These listed variables are preselected by default, but they can be removed from this list and/or other variables can be selected simply by clicking on the black arrows next to each output variable category in the center of the *Edit Field Output Request* submenu window and toggling on or off the required variable.

Press OK to confirm the selection.

History Output Requests may be reviewed in a similar way to that presented for *Field Output Requests.*

What is the difference between *Field Output* and *History Output*?

The main difference is in the way the data are collected. *Field Output Requests* contains parameters for a whole model or a larger element of the model and these data can be calculated, collected, and displayed in the form of graphs, deformed shapes, contours, or symbols (e.g., every five increments for a whole model), so calculation of the data can be performed for a whole model and all its elements, but it should also be performed in a specific time range. The *History Output* contains parameters that can be calculated, collected, and displayed in the form of graphs, rather for single nodes, to present changes of a specific parameter for this single node over time. The icon representing *Create History Output* tool is situated under the icon representing *Create Field Output* on the left side of the viewport.

2.1.7 BCs and Loads

- BCs are restrictions, usually in the form of equations, that limit the possible solutions to a differential equation. In the case of analysis performed in Abaqus, BCs are applied to those regions of the model where the displacements and/or rotations are known. There are different possibilities for restricting the regions in models prepared in Abaqus (e.g., a full restriction), which means zero displacement and rotation during the simulation, which is represented by the *Encastre* option in Abaqus.

In order to define the *BCs*, one should go to the *Model tree* on the left side of the screen and select BCs and double-click on it or select *BC* and then *Create* from the top toolbar or select an icon representing the *Create Boundary Condition* tool. As soon as one of these options is selected, a submenu window having the same name as the icon appears, as presented in Figure 2.15. This may be named the BC.

As presented in Figure 2.15, the Initial step was selected to impose the BC, which means that this specific BC will be activated in this step. The imposed mechanical BCs should have zero magnitudes in the Initial step.

If the second step called Compression is selected, the BC magnitudes can be introduced.

Depending on the modeling needs, different BCs can be imposed. If the Electrical/Magnetic or Other category of BCs is selected, other options than in the case of the Mechanical category are available for selection (e.g., for the Electrical/Magnetic category), only one type, namely Electric potential, appears.

In the case of the Mechanical category, for a step different from the Initial one, the BCs selection list is slightly different from that in the case of the Compression step, and these restrictions are as follows:

a. Symmetry/Antisymmetry/Encastre,
b. Displacement/Rotation,
c. Velocity/Angular velocity,
d. Connector displacement,
e. Connector velocity.

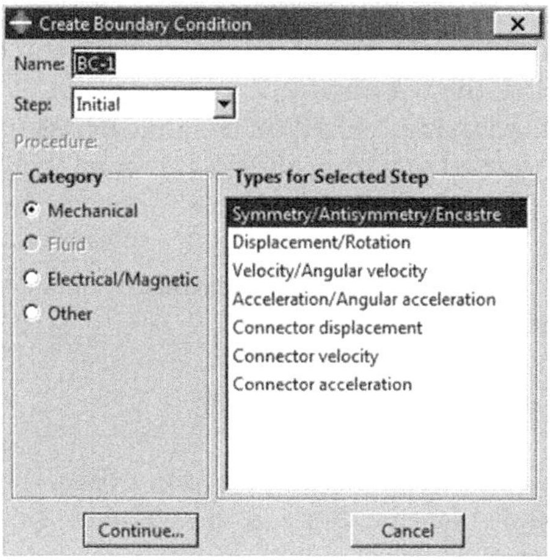

FIGURE 2.15
An exemplary submenu window for *Create Boundary Condition*. © Dassault Systèmes, a French "société européenne" (Versailles Commercial Register # B 322 306 440), or its subsidiaries in the U.S. and/or other countries.

As soon as the type of BC is selected, the user is also asked to *Select regions* for, which means marking the region of the model in the viewport where this BC is to be imposed. The selection should be confirmed by pressing the *Done* button in the prompt area. This confirmation evokes the opening of a new submenu window, examples of which are presented in Figure 2.16a–c, where

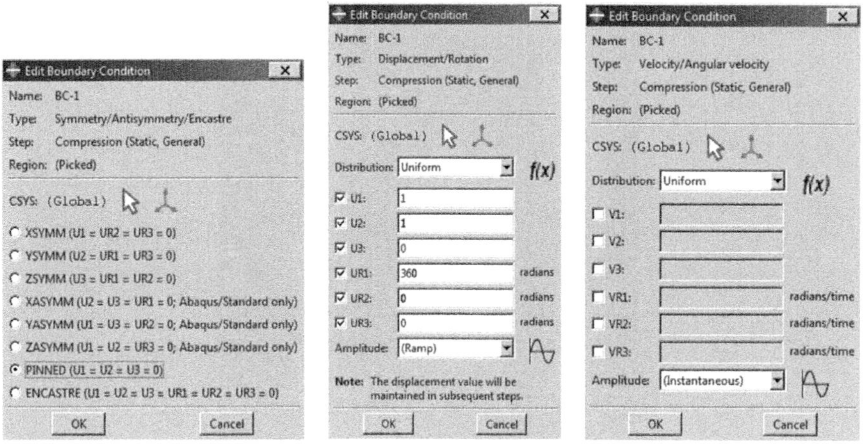

(a) (b) (c)

FIGURE 2.16
Exemplary submenu windows of editing BCs. © Dassault Systèmes, a French "société européenne" (Versailles Commercial Register # B 322 306 440), or its subsidiaries in the U.S. and/or other countries.

one can select the magnitude of the BC and also along which axes the BC is applied. The type of submenu window here depends on the preselection.

Press OK to confirm the selection.

In the case presented in Figure 2.17, on the edge of one of the components of a yarn, a BC called Pinned was imposed. This means that this edge cannot be displaced either in the direction of the *x*-axis, or in the direction of the *y*-axis and *z*-axis, which is noted symbolically as $U1 = U2 = U3 = 0$ in Figure 2.16a.

Abaqus CAE displays markers on the surface where the BC was imposed to indicate the constraints.

- Applying a load to the model

A load in Abaqus CAE refers to any type of imposed force, temperature, or any other factor that evokes a change in the response of a structure from its initial position or state.

In order to create a load go to the *Model tree* on the left-hand side of the screen and double-click on *Loads* or select Create Load icon from a toolset on the left-hand side of the viewport, or select *Load* and then *Create* from the top toolbar menu. As soon as one of these options is selected, a submenu window called *Create Load* appears, as presented in Figure 2.18. Name the file and select the step in which the load is to be applied, the category of the load and its type.

Examples of these factors are concentrated force, pressure, gravity for mechanical category (Figure 2.18), concentrated charge, surface charge, and body charge for electrical/magnetic category.

At the bottom of the Create Load submenu window, press *Continue...* button, then click Create. This action results in the appearance of the prompt

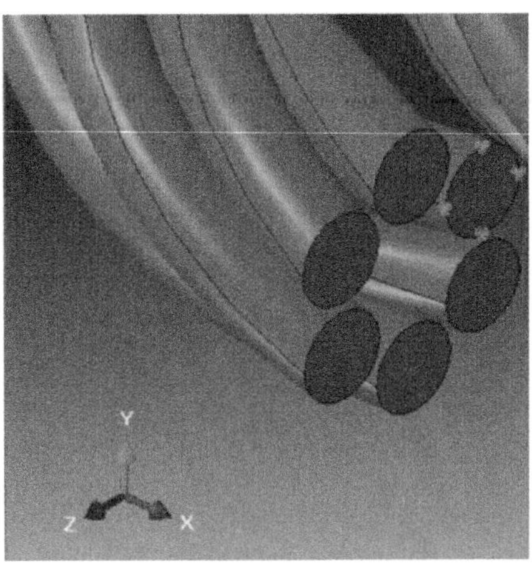

FIGURE 2.17
An example of the yarn assembly with Pinned BCs imposed at the edge of one of the components of the yarn. © Dassault Systèmes, a French "société européenne" (Versailles Commercial Register # B 322 306 440), or its subsidiaries in the U.S. and/or other countries.

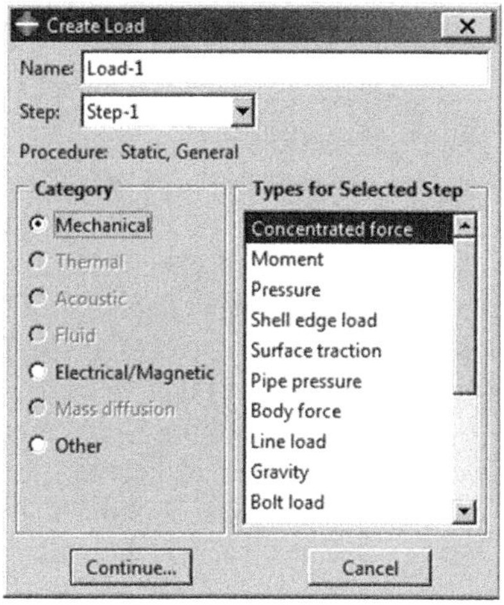

FIGURE 2.18
Example of submenu window *Create Load*. © Dassault Systèmes, a French "société européenne" (Versailles Commercial Register # B 322 306 440), or its subsidiaries in the U.S. and/or other countries.

in the prompt area, informing about the necessity of selecting a place in the model where the load is to be applied (e.g., a surface (*Select surfaces for the load*) or a point (*Select points for the load*)). After selecting the appropriate element on the model and pressing the *Done* button in the prompt area to confirm selection and execute the action, a new submenu window *Edit Load* appears, which is the place where one may define the load in terms of its vector (e.g., clicking OK in the submenu window finalizes this step). The created load will be presented symbolically in the model by an arrow with its vertex directed along the direction of the applied load.

2.1.8 Meshing and Element Types

- Meshing in finite element analysis requires dividing a model into smaller elements (finite elements), so further processing of a model is possible and analytically effective. Thanks to the meshing operation one acquires information concerning the details of the model, which sometimes has a complicated geometry, so without this division a deeper analysis of the phenomena around the model, or of situations to which the model is subject, would not be possible. Meshing also allows focusing on selected elements to operate with them instead of operating with a large body. This subchapter consists of creating a mesh (net) of elements on the existing assembly (object that one plans to model). In other words, the mesh represents a geometric object as a set of finite elements.

As presented in Figure 2.19, meshing can be straightforward, especially if it is a whole model with a relatively simple shape. It is a more complicated operation if the geometry of the model is complex. In this situation selection of the adequate mesh is a process that can be supported by Abaqus CAE.

In order to mesh a model, one should go to the *Model tree* on the left-hand side of the screen, go back to *Parts*, and select the Part which was already created in the initial step. Double-click on *Mesh (Empty)*, which is the last option in the Part branch of the *Model tree*. This action changes the color of the model from gray to yellow. In order to mesh a model, one may select the *Seed Part* icon from a toolset on the left-hand side of the viewport, or select *Seed*, and then *Part* from the top toolbar menu. As soon as one of these options is selected, a submenu window called *Global Seeds* appears, as presented in Figure 2.20. The most important aspect in this window is a parameter called *Approximate global size*.

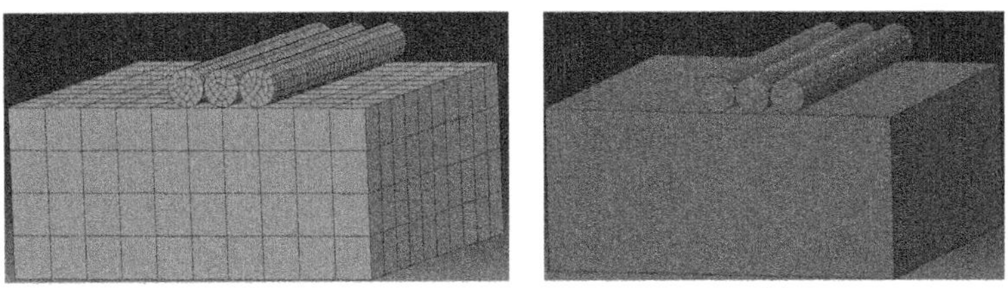

(a) (b)

FIGURE 2.19
(a) Assembly of a cubic-like elements together with three cylinders, and (b) structured coarse mesh created with an eight-node linear brick, being a regular hexahedron for a cubic-like element, and an eight-node linear brick being an irregular hexahedron for cylinders on the base of assembly (a). © Dassault Systèmes, a French "société européenne" (Versailles Commercial Register # B 322 306 440), or its subsidiaries in the U.S. and/or other countries.

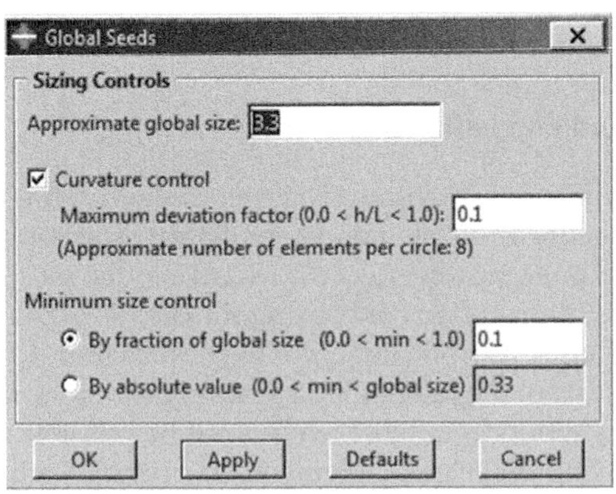

FIGURE 2.20
Example of submenu window *Global Seeds*. © Dassault Systèmes, a French "société européenne" (Versailles Commercial Register # B 322 306 440), or its subsidiaries in the U.S. and/or other countries.

The Approximate global size value in this example is 3.3, which is a default size of the element assessed by the software, based on the size of the element, which was proposed in the example. This value can be changed at any time after accessing the Global Seeds submenu. The curvature control option is toggled automatically and means that any curved elements of the model will be meshed with caution. In order to control the process, one may select a different parameter of maximum allowed deviation factor other than 0.1, presented in the example in Figure 2.20. The deviation factor is a measure of how much the element edges deviate from the original geometry.

The impact of the *Curvature control* parameter on the mesh is presented in Figure 2.21, where a simple tube is meshed with and without curvature.

As soon as seeding is established by selecting the OK button in the Global Seed submenu window, one may start selection of the meshing parameters.

Let us focus on meshing itself and continue with discretization. Different element types will be discussed in the following paragraph.

In order to *Mesh* a model, one should select Mesh from the top toolbar menu, and next *Element Type,* or select the Assign Element Type icon from a toolset on the left-hand side of the viewport. As soon as one of these options is selected, a prompt saying *Select the regions to be assigned element types* appears in the prompt area. The user is required to do this using the cursor of the mouse, and next the user needs to select *Done* in the prompt area. A submenu window called Element type appears, as presented in Figure 2.22.

Let us suppose that one is modeling an object that is to be subject to mechanical deformation. In order to mesh this object, the user should select the following parameters: Standard, 3D Stress, and Linear.

A term Standard refers here to Abaqus/Standard. Sometimes, the term "Implicit" is utilized in the literature to refer to Abaqus/Standard, which is one of the main parts of Abaqus software. Multipurpose Abaqus/Standard is meant for the performance of different types of analysis, which can deal with both linear and nonlinear problems involving static and dynamic, as well as thermal, electrical, and electromagnetic components. Abaqus/Standard approaches a system of equations characterizing models via incrementation. In contrast, Abaqus/Explicit works on solutions for specific models through time in small time increments, without solving a coupled system of equations at each increment, as takes place in the case of Abaqus/Standard. Abaqus/Explicit is meant for limited, very specific analysis that requires the utilization of a dynamic finite element approach. It is meant rather for assessment and modeling transient dynamic phenomena accruing in the modeled objects, such as impact, and is also useful for highly nonlinear problems involving changing contact conditions, such as forming simulations.

3D Stress simply stands for three-dimensional analysis of stress–strain type. Depending on the needs, a user can select Acoustic type or Cohesive, Continuum Shell, Heat transfer, etc. The selection of the type of so-called Family, as presented in Figure 2.22, determines the selection of element types.

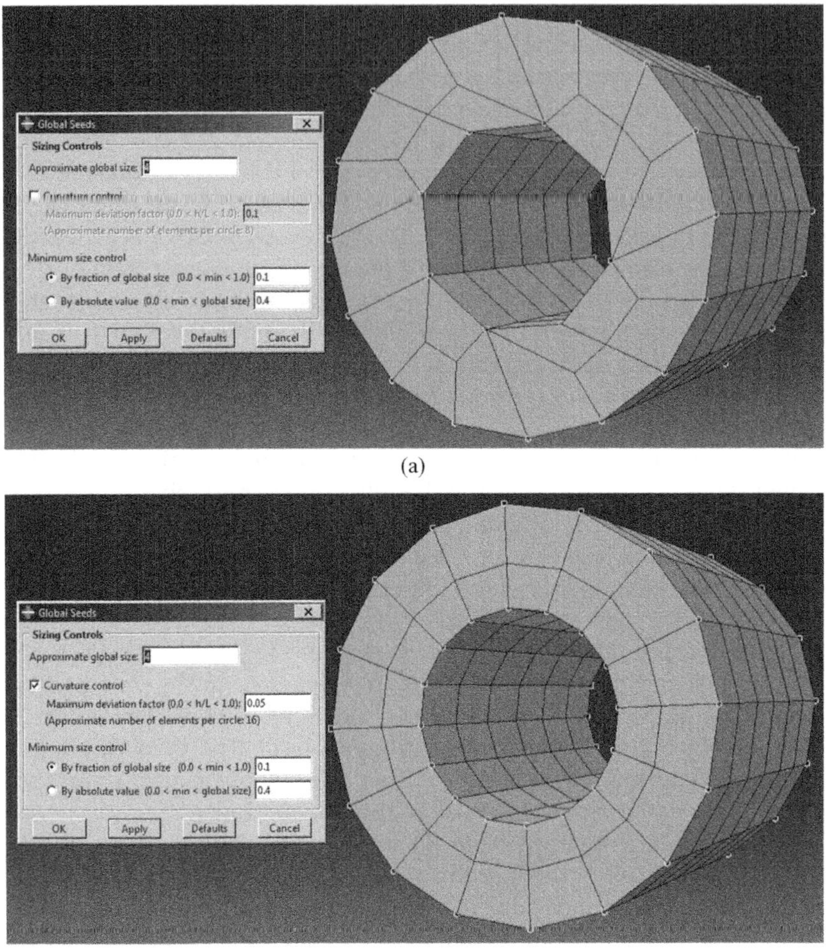

(a)

(b)

FIGURE 2.21
Global Seeds submenu together with examples of impact of curvature control parameters on the mesh quality; (a) mesh without any curvature control; and (b) mesh with a curvature control parameter on the level of 0.05. © Dassault Systèmes, a French "société européenne" (Versailles Commercial Register # B 322 306 440), or its subsidiaries in the U.S. and/or other countries.

Confirmation of the selection, as presented in Figure 2.22, allows progression to the next step, which is dividing the model into elements. The user needs to select *Mesh/Part*, which can be found in the top toolbar menu. The prompt area presents the questions: "OK to mesh the part? Yes or No?" By confirming, the user confirms all previous selections related to seeding and element type.

There are several tools in Abaqus allowing creation, meshing, and remeshing, verifying a mesh of the object to elaborate the best possible mesh.

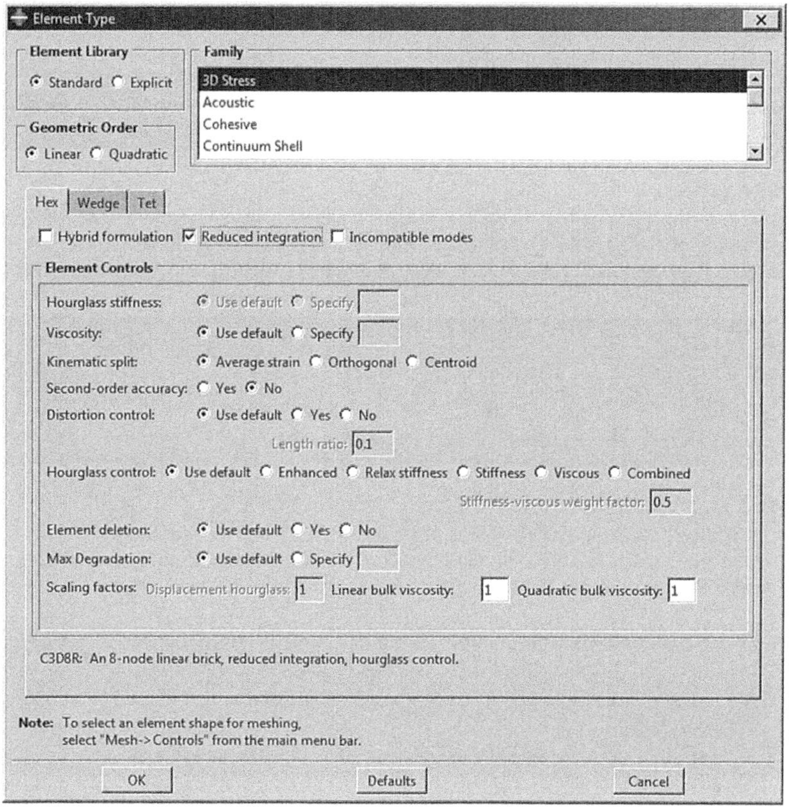

FIGURE 2.22
Example of submenu window *Element type*. © Dassault Systèmes, a French "société européenne" (Versailles Commercial Register # B 322 306 440), or its subsidiaries in the U.S. and/or other countries.

The following example presents a three-component model of the yarn, which was created in *Solid Works 2015*. The file with the model was imported into Abaqus by *File/Import/Part*.

The file model created in SolidWorks software had an IGES extension, which is compatible with Abaqus. IGES stands for Initial Graphics Exchange Specification.

After selecting the appropriate file, the submenu window called *Create Part form IGES File* appears, and some adjustments can be made. Confirming the selection with "OK" uploads the file into Abaqus software. The meshing stage of this three-component model of yarn can have different scenarios, as presented in Figures 2.23–2.25.

As can be seen, the coarse mesh was changed into a fine one, which will support the convergence of the model. Remeshing from a global mesh size of 7.6 into 1 increases the quantity of the elements, and in this specific case, it is

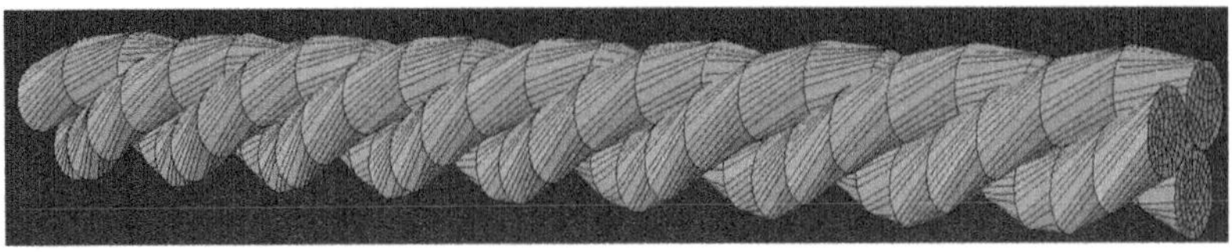

FIGURE 2.23
Exemplary mesh: initial software proposition of mesh of a three-component model of yarn; approximate global size is 7.6. © Dassault Systèmes, a French "société européenne" (Versailles Commercial Register # B 322 306 440), or its subsidiaries in the U.S. and/or other countries.

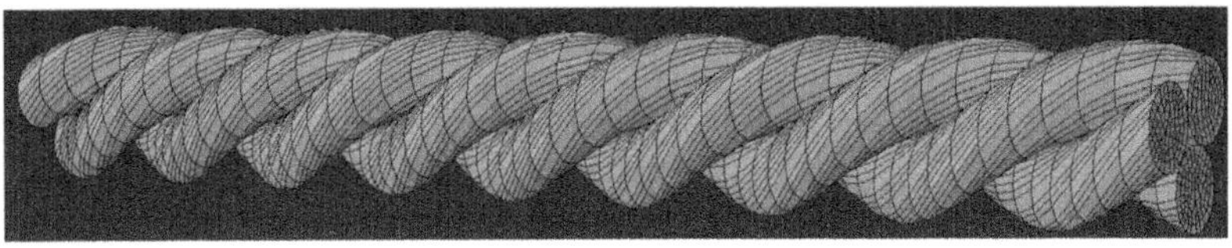

FIGURE 2.24
Exemplary mesh: customized proposition of mesh for a three-component model of yarn; approximate global size is 3.8. © Dassault Systèmes, a French "société européenne" (Versailles Commercial Register # B 322 306 440), or its subsidiaries in the U.S. and/or other countries.

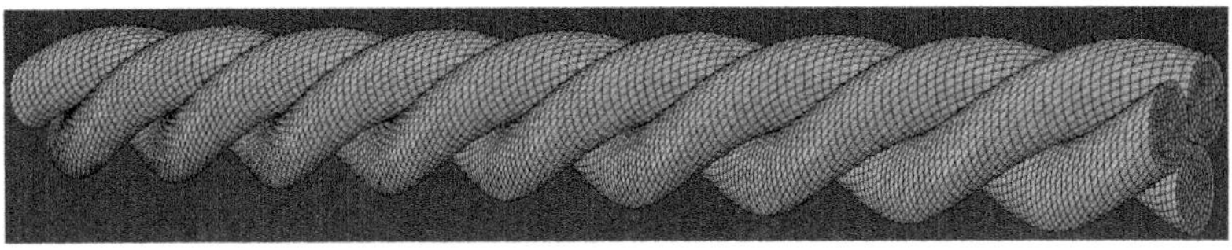

FIGURE 2.25
Exemplary mesh: customized proposition of mesh for a three-component model of yarn; approximate global size is 1. © Dassault Systèmes, a French "société européenne" (Versailles Commercial Register # B 322 306 440), or its subsidiaries in the U.S. and/or other countries.

a change from 5,750 elements into 73,863 elements. All elements are the same, namely C3D8R, which stands for a three-dimensional, continuum stress–displacement analysis of elements which are eight-node linear, with brick reduced-integration and hourglass control. In spite of the fact that the mesh presented in Figure 2.25 may seem relatively accurate, its quality should be verified by the selection of the appropriate icon, which is on the left-hand side of the viewport, or by selecting *Mesh/Verify* from the top toolbar menu.

The comments in the prompt area say *"Select the regions to verify by"* and the user may select whether it is a part, geometric region, or an element. In the presented case, one should select part (all three elements in the yarn) or "all geometric regions" (faces, edges). An element is a single fragment of the meshed object. The selection of a mode (a part or geometric regions) of the model needs to be confirmed, both in the viewport and under it in the prompt area by clicking on the *Done* button. As a consequence a new sub-menu window appears, which is presented in Figure 2.26.

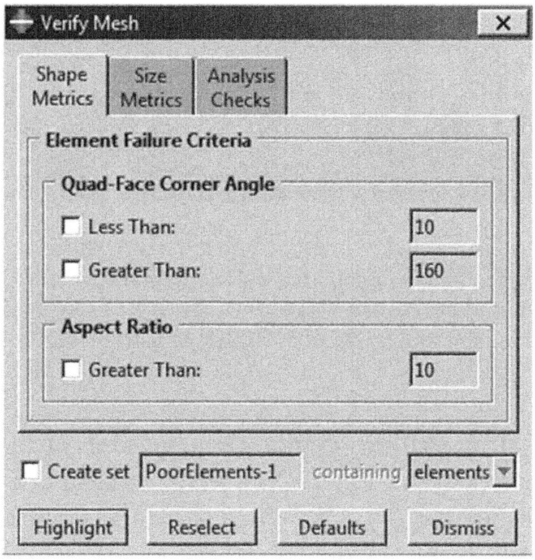

FIGURE 2.26
Example submenu window *Verify Mesh* with *Defaults* settings in a *Shape Metrics* tab for three-componential yarn model. © Dassault Systèmes, a French "société européenne" (Versailles Commercial Register # B 322 306 440), or its subsidiaries in the U.S. and/or other countries.

This menu allows the definition of a few parameters, according to which the mesh is to be verified.

In the first tab of the window, *Shape Metrics* allows the selection of the following parameters:

a. Quad-Face Corner Angle

   This allows the specification of the shape factor criterion for elements composing the mesh in the model, and in the currently presented case, there are three components of the yarn. In the case of the current model, quadrilaterally faced hexahedron elements were selected to build the mesh, thus one is able to specify the border dimensions of the faces of the elements, which are a small face corner angle and a large face corner angle. Toggling on the option Less Than and selecting five will highlight all the elements in the model that do not meet these criteria, as presented in Figure 2.27.

The selected criterion is applied for each element of the selected part in the model. After selecting *Highlight*, all the elements that do not meet the criterion are highlighted, as presented in Figure 2.27.

The term Small face corner angle relates to the elements containing faces where two edges meet at an angle smaller than a specified angle.

The term Large face corner angle relates to the elements containing faces where two edges meet at an angle larger than a specified angle.

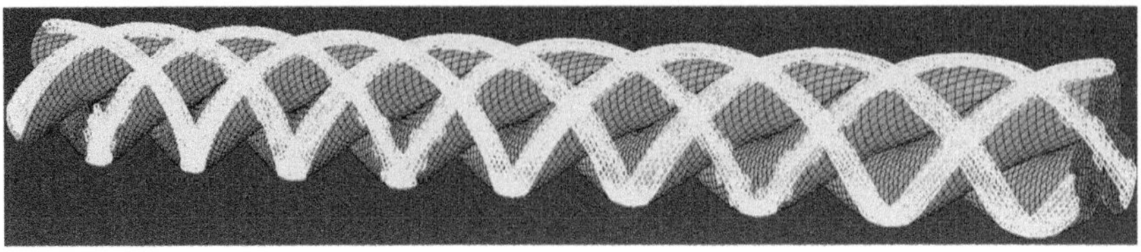

FIGURE 2.27
Model of yarn with highlighted elements that do not meet the set criterion; that is, small face element corner angle should not be lower than 5. © Dassault Systèmes, a French "société européenne" (Versailles Commercial Register # B 322 306 440), or its subsidiaries in the U.S. and/or other countries.

There is also another criterion called *Shape factor*, but it is only available for triangular and tetrahedral elements, and indicates the optimal element shape. Its value varies from 0 to 1, where 0 is a degenerate element shape and 1 is an optimal element shape.

In order to calculate a shape factor for triangular elements in the model, the user should utilize the ratio of an element area existing in the model, and an optimal element area which is the area of an equilateral triangle with the same circumradius as the element existing in the model.

To calculate a shape factor for tetrahedral elements in the model, the user should utilize the ratio of an element volume existing in the model and an optimal element volume, which is the volume of an equilateral tetrahedron with the same circumradius as the element existing in the model.

b. Aspect ratio

   This is the ratio of the longest dimension to the shortest dimension of the quadrilateral element. If the ratio increases, the inaccuracy of the solution increases. The reason for this is the fact that elements yield the best results if their shape is regular, and regularity requires the exact or similar length of the edges of elements.

In the case of *Size Metrics* tab, the user may define the following parameters supporting the process of creating a more accurate mesh, which are presented in Figure 2.28:

a. Geometric deviation factor

   The geometric deviation factor is a measure of how much an element edge deviates from the original geometry. It can be calculated by dividing the maximum gap between an element edge and its parent geometric face or edge by the length of the element edge. By default, the software highlights the elements that have a geometric deviation factor greater than 0.2, and this default value is presented in Figure 2.28.
b. Edge shorter than and edge longer than are two parameters imposing limits on the dimensions of the element. The imposed limitations aim

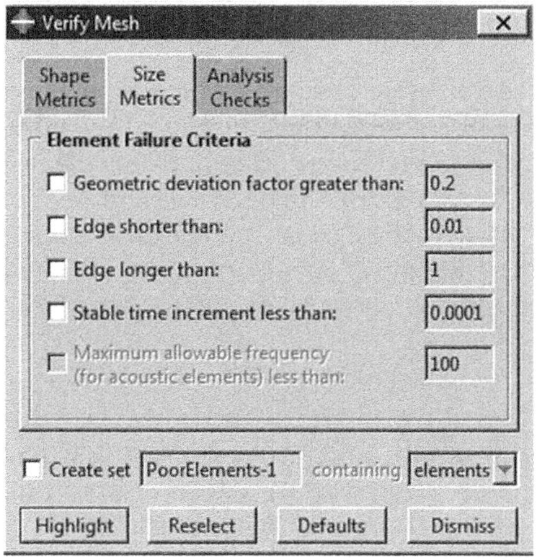

FIGURE 2.28
Example of submenu window *Verify Mesh* with *Defaults* settings in a *Size Metrics* tab for three-componential yarn model.

to maintain the more regular shape of the elements in the model. If they are shorter or longer than the lengths specify in the Size Metric tab menu (the default values are presented in Figure 2.28.), the software will highlight the elements that exceed the maximum allowed length.

c. Stable time increment

The calculated stable time increment is less than the specified value, which is 0.0001 in the examples, presented in Figure 2.28. The full stability depends on the material properties ascribed to the model.

The third tab called *Analysis Checks* allows to toggle on/off two options, namely Errors and Warnings. If any of the conditions listed in both the Shape Metrics and Size Metrics tabs are not fulfilled, and these two parameters are toggled on, the specific element of the model, which does not meet the requirements, will be highlighted, as presented in Figure 2.27, where only Warnings are highlighted as there are no errors in the model. This means that all the elements of the model meet the requirements in terms of their shape and size, but some of them are very close to the given limit. Additionally, the user may find some precise information in the prompt area under the model (e.g., Number of elements 73,863) which refers to the analysis of all elements in the model. Analysis errors: 0 (0%); there are no errors in the model in terms of the elements size and shapes; and none of the given parameters, being restrictions of dimensions of the elements, were exceeded; Analysis warnings: 4,301 (5.82294%); the software highlighted 4,301 elements, which is 5.82% of the total number of elements and is very close to the given dimensional or shape limit.

In order to remove warning or errors, the user should improve/refine the mesh.

How to discretize and select the appropriate element?

The total number of elements that the model is composed of as well as its size and type depend on the user. There is no rule, which says that for a specific model, 500k hexahedral elements is right or wrong. A responsible and rational user with some engineering abilities and some adequate experiences supported by literature review on a modeled subjects is to decide. There are some basic tips on how to discretize the object:

- It should be composed of elements, which are small enough to provide with a useful set of data; a good practice is to start with what Abaqus CAE prompts, and then compare that with three or four attempts (or more if necessary), which change the size of the elements. The results of all the attempts should be compared. If the results provided are not changed, it does not make sense to divide the element anymore.
- The element size needs to be large enough so as not to affect the computational costs.
- Reduce the element size if the modeled phenomena or object are exposed to some kind of drastic change or, for example, only a small part of the model is subject to a larger deformation, so this smaller region should have a finer mesh. A great example of a composite subject to deformation is that of a penetrating bullet: the region where the bullet impacts should have a finer mesh than the rest of the composite.

The user may encounter the term "an orphan mesh" when elaborating models. An orphan mesh is a mesh that has no geometry associated with it. The orphan mesh can be obtained/created in the following ways:

- By importing an STL file into Abaqus and using Plug-ins/STL Import. This allows the user to import only the nodes/elements.
- By importing an Initial Graphics Exchange Specification (IGS or IGES) file (*.igs or *.iges) into Abaqus and using Import/Part. As soon as IGS file exists as a meshed Part in the *Model tree*, the user should go to Mesh/Create Mesh Part, which is in the top toolbar menu. A confirmation of the name selection of the existing file by pressing OK in the prompt area generates an orphan mesh based on this existing file.

Part of the model with an orphan mesh can be combined/contacted with other parts that have a normal mesh parts in the frame of one model.

- Element type

The elements of which the discretized model consists may have different forms depending on what kind of shape was discretized and what kind of model was discretized (e.g., a solid body or truss). They also depend on the particular interest of the software user and the specific needs, requirements that the model needs to meet, or the modeled phenomenon itself.

Existing classifications presented in the literature have different accuracy and order of presenting elements (e.g., Getting Started with Abaqus: Interactive Edition for Abaqus 6.13, 2016 Dassault Systèmes; Logan, 2012).

There are different approaches according to which single element of the model can be classified (e.g., elements can be classified according to the family of elements, their degrees of freedom, number of nodes existing in the element, formulation of the element or their integration).

The user encounters the term Element type when meshing the model and selecting an adequate type of the smallest components (elements) of which the whole model is built.

As presented in Figure 2.22, an element is an important fragment of the whole model, and there are a few parameters that need to be ascribed to the element to satisfy the needs of a correct modeling.

An element type submenu window presented in Figure 2.22 allows the selection of several parameters. The differences between the Standard and Explicit versions of Abaqus CAE have already been explained. The user may also make a choice between *Linear* and *Quadratic geometry order*. In short, the linear elements utilize a linear approximation of displacement field over the domain of the element, and Quadratic elements utilize a quadratic approximation of the displacement field. Making a choice between Linear and Quadratic elements impacts the possible selection of the shape of the 3D elements and their characteristics (e.g., if the users select Linear geometric order, they can choose among *Hex elements* (with Hybrid formulation or with Reduced integration or with Incompatible modes), *Wedge elements* (with Hybrid formulation), or *Tet elements* (with Hybrid formulation)). And if the users select Quadratic geometric order, they can choose among Hex elements (with Hybrid formulation or with Reduced integration), edge elements (with Hybrid formulation), or Tet elements (with Hybrid formulation or with Modified formulation or with Improved surface stress formulation).

Hybrid formulation is utilized when the material characteristics ascribed to the material of the model are incompressible (Poisson's ratio = 0.5) or very close to incompressible. Thus, the hybrid formulation option of the elements in the model can be utilized when the user is modeling human skin or other soft biological materials, stretchable materials (e.g., fabrics with polyurethane).

As their name suggests, *Reduced-integration elements* utilize fewer integration points to represent elements. In fact, they utilize one less integration point in each direction than the fully integrated elements.

Incompatible mode can be selected if the user is sure that some shear problems with materials are expected. Thus, a shear-locking option is applied in, for example, a situation of the element's displacement field's inability to model bending. More details and graphical examples are provided in the literature (Getting Started with Abaqus: Interactive Edition for Abaqus 6.13, 2016 Dassault Systèmes; Logan, 2012).

In the case of some 2D and 3D models, for instance, a reduced-integration mode for a linear geometry order engages *hourglass* deformation of the element. The term "hourglass" refers to the deformation of a mesh of elements, being a direct consequence of the excitation of zero-energy degrees of freedom. In a deformed model, an hourglass option of elements is presented as deformed hourglass-like element shapes, where individual elements are severely deformed, whereas the overall mesh section is not deformed.

2.1.9 Job

To create a Job, the user should go to the *Model tree* on the left-hand side of the screen and double-click on *Job*, select *Create Job* icon from a toolset on the left-hand side of the viewport, or select Job and then Create from the top toolbar menu. As soon as one of these options is selected, a submenu window called *Create Job* appears. The user may name the file, and the user should make sure that the adequate model selection using the *Continue* key evokes the appearance of another submenu window called Edit Job, as presented in Figure 2.29.

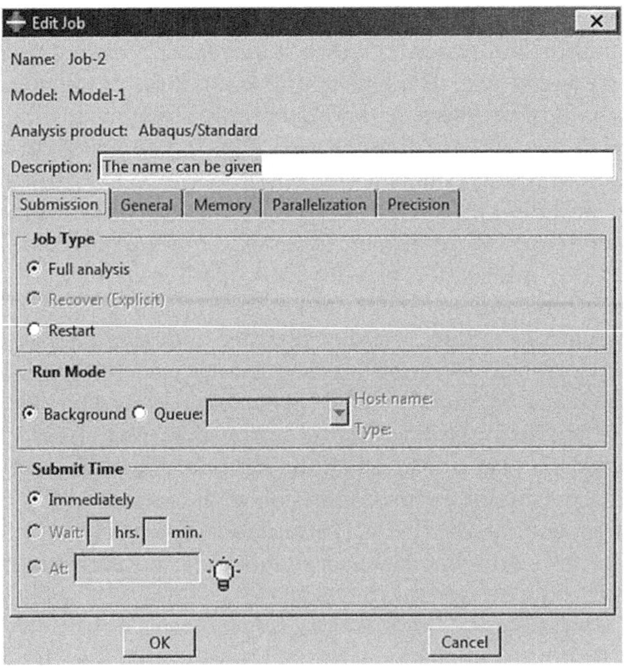

FIGURE 2.29
Example of submenu window called *Edit Job*. © Dassault Systèmes, a French "société européenne" (Versailles Commercial Register # B 322 306 440), or its subsidiaries in the U.S. and/or other countries.

In fact, the best option for the user is to rely on the default selection offered by Abaqus. Thus, it is advised to select the OK button to confirm the step. The Edit Job submenu allows the choice of model-data processing and recording, but not the essential outcome of the analyzed object or the phenomena in the created model.

Next, select the created Job (Job-2, as presented in Figure 2.29) in the *Model tree* by double-clicking on the Jobs branch, as presented in Figure 2.30.

The user should first check the setting and then submit them for analysis performed by Abaqus. This can be done by clicking on Job-2 using the left-hand mouse button and selecting the Data Check option. It can also be accomplished by selecting Job and the appropriate options in the top toolbar menu. Depending on the model, Data Check may be a very fast process or it may take some time. At the same time, depending on the Data Check stage, a prompt area presents the progress of Data Check by presenting comments gradually (e.g., The job input file "Job-2.inp" has been submitted for analysis; Job Job-2: Analysis Input File Processor completed successfully; Job Job-2: Abaqus/Standard completed successfully; Job Job-2 completed successfully).

As soon as these comments appear, the Job may be submitted. This can be done by clicking on Job-2 using the left-hand mouse button and selecting Submit. The software allows the monitoring of the processing of the model's submitted data while it runs the model. In fact, the option Monitor presents a large and useful dataset (e.g., Error reports, Warning reports (if any, they will be listed there), Output file name, Data file containing the whole script written in Python scripting language, the characteristic of the created model (number of elements, number of nodes, total number of variables in the model, etc.), job summary, and step summary).

If the model was created correctly and the software did not encounter any issues while processing, the same set of comments as in the case of Data Check will be displayed in the prompt area, and are also available in the Monitor submenu window. The job may be aborted if the software encounters a problem with the input file.

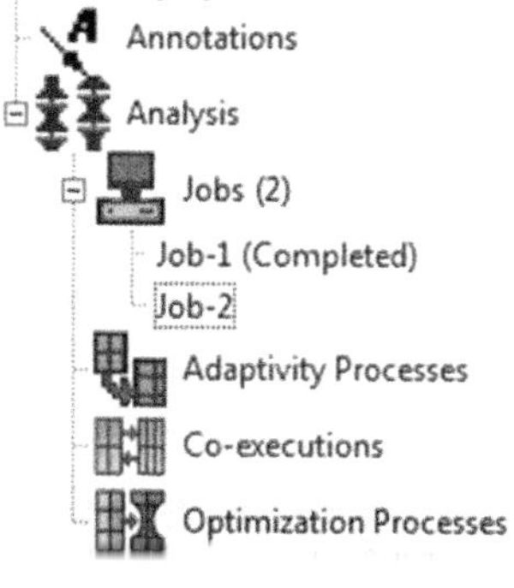

FIGURE 2.30
Fragment of *Model tree* showing Jobs branch, with two different jobs created. © Dassault Systèmes, a French "société européenne" (Versailles Commercial Register # B 322 306 440), or its subsidiaries in the U.S. and/or other countries.

2.1.10 Solution, Visualization

To observe and verify the achieved solution for the model, the user should go to Job in the *Model tree* and select Job-2 using the left-hand mouse button and selecting the Results option. This switches automatically into the software Results mode. This can be also achieved by selecting Job/Results/Job-2 in the top toolbar menu.

The fact of the software switching into the Result menu is shown by three major visuals: the Results *Model Tree* is present in a separate tab on the left-hand side of the screen; a set of tools supporting analysis; and a viewport with processed model, as presented in Figure 2.31. Additionally, a top toolbar is adapted to the Results.

A set of the following tools, set of tools no. 3, appears on the left-hand side of the viewport with the model, supports the visualization of the model and the visualization of the results, and allows the user to obtain the desired

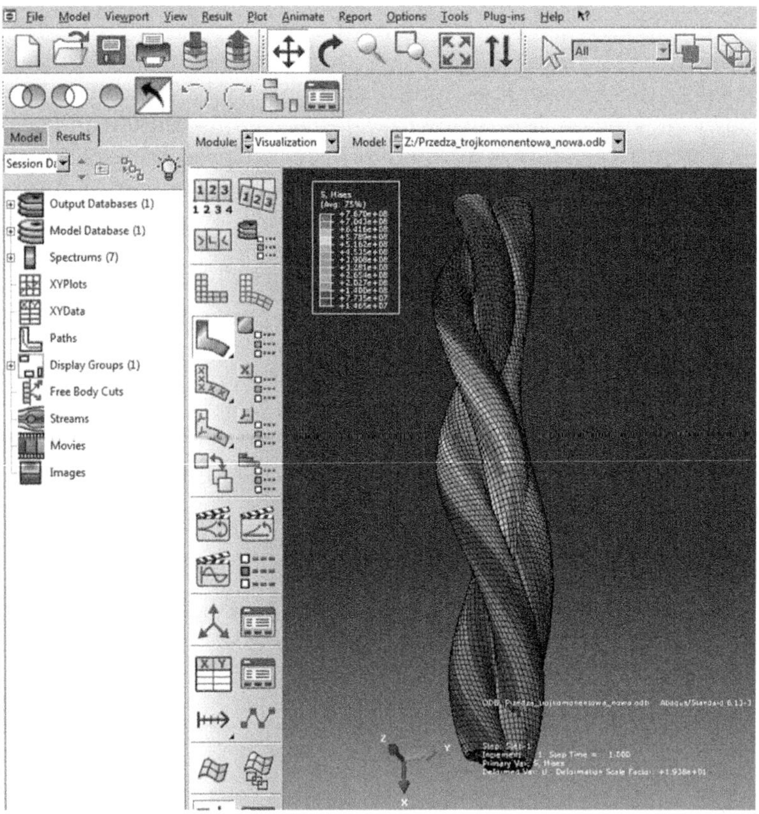

FIGURE 2.31
Example view of Results panel in Abaqus CAE. © Dassault Systèmes, a French "société européenne" (Versailles Commercial Register # B 322 306 440), or its subsidiaries in the U.S. and/or other countries.

deformed or undeformed view/cut for the specifically measured parameter selected in the Field Output Request and History Output Request:

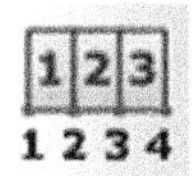 Icon representing *Common Plot Options* tool (e.g., style, colors, labels of plots)

 Icon representing *Superimposed Plot Options* tool

 Icon representing *Result Options* tool

 Icon representing *Odb Display Options* tool

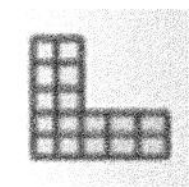 Icon representing *Plot Undeformed Shape* tool

 Icon representing *Plot Deformed Shape* tool

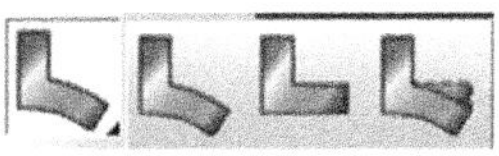 A group of icons representing *Plot Contours on Deformed Shape*, *Plot Contours on Undeformed Shape*, and *Plot Contours on both shapes* tools, respectively

 Icon representing *Contour Options* tool

 A group of icons representing *Plot Symbols on Deformed Shape*, *Plot Symbols on Undeformed Shape*, and *Plot Symbols on both shapes* tools, respectively

 Icon representing *Symbol Options* tool

 A group of icons representing *Plot Material Orientation on Deformed Shape*, *Plot Material Orientation on Undeformed Shape*, and *Plot Material Orientation on both shapes* tools, respectively

 Icon representing *Material Orientation Options* tool

Icon representing *Allow Multiple Plot States* tool

Icon representing *Ply Stack Plot Options* tool

Icon representing *Animate: Scale Factor* tool

Icon representing *Animate: Time History* tool

Icon representing *Animate: Harmonic* tool

Icon representing *Animation Options* tool

Icons representing *Create Coordinate System* and *Coordinate System Manager* tools, respectively

Icons representing *Create XY Data* and *XY Data Manager* tools, respectively

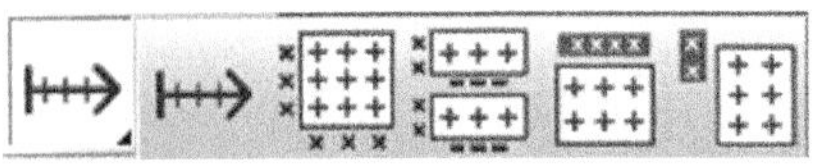

Icons representing *XY Axis Options, XY Chart Options, XY Plot Options, XY Plot Title Options,* and *XY Legend Options* tools, respectively

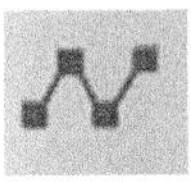

Icon representing *XY Curve Options* tool

Icon representing *Create Field Output From Fields* tool

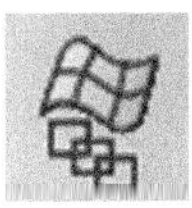

Icon representing *Create Field Output From Frames* tool

Icons representing *Activate/Deactivate View Cut* and *View Cut Manager* tools

Icons representing *Create Free Body Cut* and *Free Body Cut Manager* tools

Icons representing *Create Stream* and *Stream Manager* tools. This tool helps to analyze the velocity or vorticity data, as well as the quantity of points along which one would like to display streamlines of fluid flow data.

Icons representing *Switch Between Overlay and Single Plot State* and *Overlay Plot Layer Manager* tools

Icon representing *Probe Value* tool

Set of tools no. 3. Set of icons allowing analysis of the results. © Dassault Systèmes, a French "société européenne" (Versailles Commercial Register # B 322 306 440), or its subsidiaries in the U.S. and/or other countries.

The user is free to utilize visualization tools present around the viewport or to select them from the top toolbar. The view manipulation tools help to examine the deformed model. The user may select the view depending on the parameter predefined. Apart from the visualization of a whole model the user may also elaborate a customized report and plots.

In order to obtain a report, the user should select *Results/Field Output* from the top toolbar. This generates a Report Field Output submenu window where the user may select a set of parameters calculated by Abaqus for a whole model. The user may select the nodes, integration points, elements, etc., for which the parameters will be listed in the report.

In order to obtain a customized XY plot, the user can select an icon representing *Create XY Data,* or an icon representing *XY Data Manager* to select a set of data for the graph, and by doing so the user may obtain a graph of stress versus time for the selected elements of the model or node. The graphs obtained this way may be customized if the user selects Options/XY Plot from the top toolbar.

The following chapters present the modeling process based on different examples (e.g., fiber, yarn, or fabrics), so the user may gain a deeper understanding of the modeling process of these textile objects, and the obtained knowledge will help the user to understand and utilize the software, in general, for the modeling of textile objects. It is strongly advised to verify one's own selection made during the modeling process using different sources (e.g., Getting Started with Abaqus: Interactive Edition for Abaqus 6.13, 2016

Dassault Systèmes, or other sources released by Dassault Systèmes, as well as previously elaborated textile models, and most of all, using finite element modeling process handbooks). Some of these sources are listed in the References section at the end of each chapter.

References

Ali, Aidy, Hosseini, M., and Sahari, B.B. (2010). A review of constitutive models for rubber-like materials. *Am. J. Eng. Appl. Sci.* 3(1): 232–239.

Arruda, E.M., Boyce, M.C. (1993). A three-dimensional model for the large stretch behavior of rubber elastic materials. *J. Mech. Phys. Solids* 41(2): 389–412.

Bergström, Jörgen. (2015). *Mechanics of Solid Polymers - Theory and Computational Modeling*. Elsevier. Online version available at: http://app.knovel.com/hotlink/toc/id:kpMSPTCM06/mechanics-solid-polymers/mechanics-solid-polymers.

Fung, Y.C. (1993) *Biomechanics: Mechanical Properties of Living Tissues*. Springer, New York.

Fung, Y.C., Fronek, K., Patitucci, P. (1979, November). Pseudoelasticity of arteries and the choice of its mathematical expression. *Am. J. Physiol.*, 237(5): H620–H631.

Gasser, T.C., Ogden, R.W., and Holzapfel, G.A. (2006). Hyperelastic modelling of arterial layers with distributed collagen fibre orientations. *J. Royal Soc. Interface* 3: 15–35.

Gent, Alan N. (2012). *Engineering with Rubber - How to Design Rubber Components*, 3rd edn., Hanser Publishers. Online version available at: http://app.knovel.com/hotlink/toc/id:kpERHDRCE2/engineering-with-rubber/engineering-with-rubber.

Getting Started with Abaqus: Interactive Edition for Abaqus 6.13, (2016) Dassault Systèmes, available online.

Holzapfel, G.A., Gasser, T.C., and Ogden, R.W. (2000). A new constitutive framework for arterial wall mechanics and a comparative study of material models. *J. Elast.*, 61(2000): 1–48.

Humphrey, J.D. (1995). Mechanics of arterial wall: Review and directions. *Critical Reviews in Biomed. Engr.*, 23(1–2): 1–162.

Li, Y. and Dai, X.-Q. (2006). *Biomechanical Engineering of Textiles and Clothing*. Woodhead Publishing in Textiles, Cambridge.

Logan, D. (2012). *A First Course in the Finite Element Method*, 5th edn., Cengage Learning, Delhi.

Marlow, R.S. (2003). A general first-invariant hyperelastic constitutive model in Constitutive Models for Rubber III. *Proceedings of the Third European Conference on Constitutive Models for Rubber*, Busfield J., Muhr A. (eds.) 15–17 September, London, UK.

Marlow, R.S. (2008). A second-invariant extension of the marlow model: Representing tension and compression data exactly. *Abaqus Users' Conference*, Newport, Rhode Island, USA.

Ogden, R.W. (1972). Large deformation isotropic elasticity-on the correlation of theory and experiment for incompressible rubberlike solids. *Proc. R. Soc. Lond. A* 326: 565–584.

Rivlin, R.S. (1948). Large elastic deformations of isotropic materials. IV. Further developments of the general theory. *Philos. Trans. R. Soc. Lon. Ser. A Math. Phys. Sci.*, 241(835): 379–397. doi:10.1098/rsta.1948.0024.

Ruíz, M.J.G. and González, L.Y.S. (2006). Comparison of hyperelassstic material models in the analysis of fabrics. *Int. J. Clothing Sci. Technol.* 18(5): 314–325.

Sacks, M.S. (1999). A method for planar biaxial mechanical testing that includes in-plane shear. *ASME J. Biomech. Eng.* 121(5): 551–555. doi:10.1115/1.2835086.

Yeoh, O.H. (1993). Some forms of the strain energy function for rubber. *Rubber Chem. Technol.* 66: 754–771.

3

Modeling of a Fiber

3.1 Theoretical Approach

Let us suppose that we are attempting to assess some of the mechanical parameters of nylon fiber. Nylon is one of the most widely produced, multi-purpose fibers. The most common way to perform the assessment is to carry out a tensile test. A tensile test is a typical engineering procedure used to characterize elastic and plastic deformations related to the mechanical characteristics of materials. A tensile force can be applied to the fiber by a device such as an Instron or a Statimat. This results in the gradual elongation and, in many cases, fracture of the test item. A fiber is clamped in between the clamps of the tensile test apparatus. One of the clamps is fixed, and the other is pulled in one direction for a period of time, during which the stress (force per unit area) and strain (percentage elongation) on the fiber are measured. This allows the creation of a stress–strain curve for a tested material. Abaqus Complete Abaqus Environment (CAE) is capable of simulating a tensile test procedure (Jose and Anto, 2015). CAE is a backronym of Computer-Aided Engineering. For simplicity and transparency of the modeling process, we will apply stresses that remain in the elastic region of stress–strain dependence for a nylon fiber. The elastic region, that is the region over which the fiber will return to its original length once the stress is removed, corresponds to the linear part of Hooke's law. Nowadays, the mechanical characteristics of materials, including raw materials used for textiles, are well documented. Some examples are shown in Figure 3.1.

The literature on this subject provides us with precise information on the parameters of most of the widely used materials depending on their final application. This includes textile materials such as different nylon fibers. Detailed characteristics of nylon and other fibers and bulk materials are listed in Table 3.1.

To model any given material, the mechanical parameters such as Young's modulus and Poisson's ratio are required. Thus, before starting modeling, it is worth performing a literature survey or performing one's own tests to establish these mechanical characteristics. As nylon is a popular material, applied in ropes, nets, toothbrushes, carpets, and even as guitar strings, its

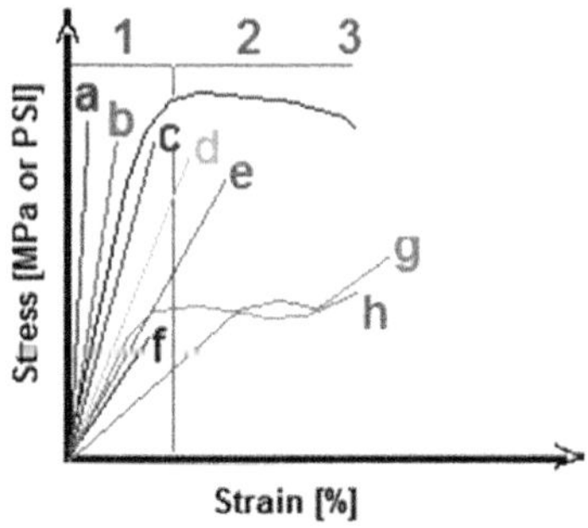

FIGURE 3.1
Simplified stress–strain characteristics of selected materials based on a review of relevant literature (Alagirusamy and Das, 2010; Bunsell, 2009; Hearle, 2001; McKeen, 2015). *1* corresponds to the elastic deformation region; *2* corresponds to the plastic deformation region; and *3* represents the point of failure. The materials or groups of materials are (a) carbon fibers, (b) Spectra 1,000, (c) Kevlar-like fibers, (d) Vecran, (e) glass fibers, (f) cotton, (g) hyperelastic materials, and (h) polyamide (nylon). Note that *g*, hyperelastic materials, is a large group of materials ranging from sponge to gels and human skin. The shape of the curve for materials in this group depends on the specific material and may vary significantly.

mechanical parameters are well known. Therefore, one has multiple sources to verify and analyze any modeling outcome.

Before describing a simulation of a tensile test using a computer program (in this case, Abaqus CAE), one important matter needs to be briefly discussed. It concerns a theoretical approach to the simulated object and the stress–strain notions in relation to the object which is to be subjected to the tension or compression.

Probably the best way to understand the basic concept here is to use an example. Consider an object, which is a cylinder or fiber-like body, and subject it to tension or compression. In reality, of course, deformations can take other forms (e.g., shear or torsion). Let us now apply a uniform static load along the axis of the cylinder. This is depicted in Figure 3.2. The cylinders are deformed due to the force (tensile load and compressive force) that is applied uniaxially along the axes of the cylinders.

To analyze the strain–stress relation for this cylinder, we start with Hooke's law. This states that the force F needed to extend or compress a body by a certain length is proportional to that length. This is true for elastic bodies or bodies, which, in the initial stage of deformation, can be characterized as elastic.

Hooke's law has the following form:

$$F = k \times \Delta L \tag{3.1}$$

where F is the force required to deform (extend it or to compress) the object by length ΔL. F is therefore proportional to ΔL. Referring to Figure 2.2, $\Delta L = L_1 - L$. k is a constant that is characteristic of the material that the object is made of. It is called *material stiffness* in the case of compression.

TABLE 3.1

Selected Bulk Materials and Fiber s and Their Characteristics (Various Sources)

Substance	Density (kg/m³)	Young's modulus (GPa)	Poisson's ratio	Source
Steel	7,480–8,000	200–209	0.3	Getting Started with Abaqus (2016), Kipp (2010), Krevelen and van Nijenhuis (2009)
Carbon Fiber (polyacrylonitrile (PAN)-based parameters)	1,770–1,900	200–250 (high-strength) 280–300 (intermediate modulus) 350–600 (high modulus)	0.1–0.33	Eichhorn et al. (2009), Donnet and Bansal (1990), Bajpai (2013), Mather, and Wardman, (2015)
Spectra-type fiber	1,135–1,320	110–113	0.402–0.5	Alagirusamy and Das (2010), Erhard (2006), Plastics Design Library (1991), Spectra characteristics (2016), Wang and Liu (2011)
Kevlar 29	1,440	70.5	0.35	Alagirusamy and Das (2010), Hearle (2001)
Kevlar 49	1,450	112.4–149	0.35	
Kevlar 149	1,470	179	0.35	
Vectran HS LCP	1,400–1,410	52–103	0.37	Hearle (2001), Kipp (2004, 2010), Plastics Design Library (1991)
C glass fiber	2,520–2,550	68.9–72	0.20–0.23	Alagirusamy and Das (2010), Mills (2005), Ross (1999)
E glass fiber	2,480	72.3	0.20–0.23	Alagirusamy and Das (2010), Ross (1999)
Polyamide	1,129–1,640	2–4	0.34–0.44	Baboian (2005), Krevelen and van Nijenhuis (2009), Rosato and Rosato (2003)
Wool	1,300–1,310	2		Bunsell (2009), Veit (2012)
Cotton	1,500	5.5–10		Belgacem and Gandini (2008)
Elastomers	1,270	0.01–0.1 0.0007–0.004	0.495–0.5	Ashby and Jones (2012), Baboian (2005), Gent (2012)
Polyester	950–1,380	15	0.36	Baboian (2005), Bunsell (2009), Kipp (2010), McKeen (2015)

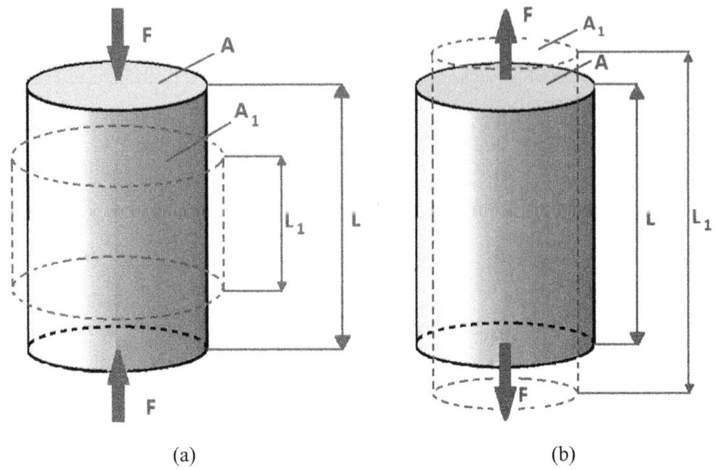

FIGURE 3.2
Deformation of cylinders under the influence of F force being (a) a compressive force and causing a contraction of a cylinder. (b) a tensile load and causing elongation of the cylinder. The dashed lines represent the cylinders after deformation; A is a cross-sectional surface of a cylinder before deformation, and A_1 is a cross-sectional surface of the cylinder after deformation; L is the length of a cylinder before deformation, and L_1 is the length of a cylinder after deformation.

The simplest example of the application of Hooke's law is a tensile test of an object, and in the current example, the object is a cylinder. The absolute deformation of this cylinder is proportional to the force F applied to the cylinder and to its length and is inversely proportional to its cross-sectional area A and to material coefficient called the *modulus of elasticity*. Besides *modulus of elasticity*, the terms *elastic modulus, tensile modulus*, and *Young's modulus* are also used for the same material coefficient. It can be seen as a value that measures an object's or substance's resistance to being deformed elastically and reversibly, when a force is applied to it.

The stress σ on the cylinder is the force per unit area $\sigma = \frac{F}{A}$ and can be measured in Pascals [Pa = N/m^2], megapascals [MPa = N/mm^2], gigapascals [GPa = kN/mm^2] or psi—pounds per square inch [1 psi = 6 894.75 Pa]. σ is given by

$$\sigma = E\varepsilon \tag{3.2}$$

where E is Young's modulus [which as a consequence of a definition presented earlier is the ratio of stress σ (which has units of pressure) to strain ε, being dimensionless, and so Young's modulus has units of pressure] and ε is the strain of the material. ε can be defined as the ratio of the increase in length of the cylinder, ΔL, to its original length, L.

$$\varepsilon = \frac{\Delta L}{L} \tag{3.3}$$

where $\Delta L = L_1 - L$. L_1 is the length after applying the stress. The absolute deformation can therefore be presented as

$$\Delta L = \frac{LF}{AE} \tag{3.4}$$

where F is the applied force, A is a cross-sectional area of the cylinder, L is the initial length of the cylinder, and E is Young's modulus.

Textbooks tend to use true stress, which is the nominal load F divided by the instantaneous cross-sectional surface of the sample (A_1), whereas in engineering terms, stress is the nominal load F divided by the initial (undeformed) cross-sectional surface of the sample (A).

Let us consider the following conditions for the tensile test of a nylon fiber with a circular cross section and with a radius of 0.5 mm or 0.0005 m. The cross-sectional area is therefore $A = 0.785\,\text{mm}^2 = 7.85\text{E}{-7}\,\text{m}^2$, and the initial length of the fiber is $L = 0.01$ m. The pressure σ applied to the surface of one of the faces of the fiber is 1 MPa (remembering that σ is the ratio of force F applied to the cross section A of the cylinder face). The fiber is composed of a polyamide with a Young's modulus of 2.5 GPa.

To calculate the deformation of the fiber, we use formula (3.4). After inputting the values above this gives $\Delta \text{L} = 0.000004$ m $= 0.004$ mm. So the change of length or absolute deformation is 0.04 mm. This means that after applying a uniform pressure to the cross section of the fiber it was elongated by 0.04 mm.

The next subchapter, where modeling of a nylon fiber is presented, allows us to verify the mathematical calculation's outcome.

3.2 Finite Element Model of a Fiber

The ten steps presented later describe the creation of an appropriate geometry, assigning material properties to this geometry, programming the activities (steps) for this geometry, setting boundary conditions (BCs) and a load, verifying the correctness of the prepared model, running the whole model, and analyzing and validating the results.

STEP 1: Creating the geometry of a nylon fiber.

- Initiate the Abaqus Standard program and, if necessary, select the icon *New* to create a new model as shown earlier. Usually the software is ready to start, so the selection of *New* is not required (Figure 3.3).
- In the *model tree*, double-click on *Parts* and select *Create*.

One may give a *Name* to the model. In case of the current model, one uses nylon fiber.

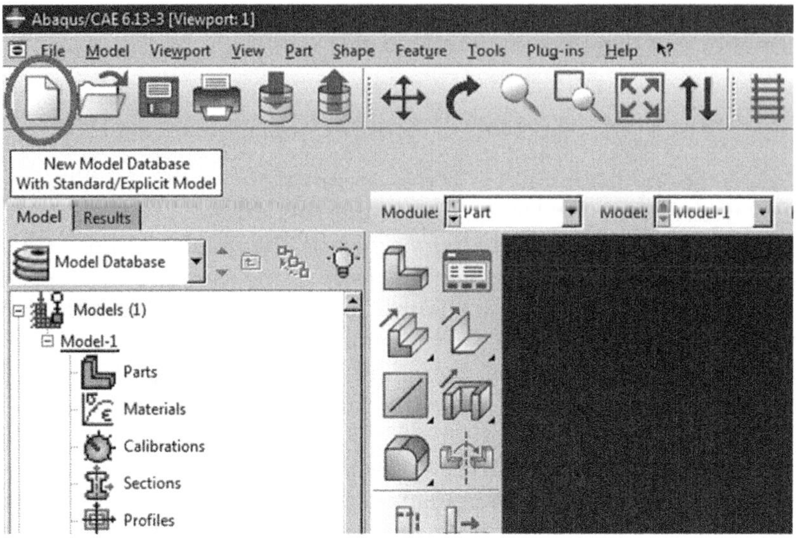

FIGURE 3.3
A fragment of an initial screen, viewport from Abaqus/CAE 6.13-3. © Dassault Systèmes, a French "société européenne" (Versailles Commercial Register # B 322 306 440), or its subsidiaries in the U.S. and/or other countries.

One models the nylon fiber as a three-dimensional (*3D*), *Deformable, Solid* body. These options should be selected as shown in Figure 3.4.

The way the model of the fiber is to be created is *Extrusion*, so highlight this option on the list.

Approximate size refers to a grid, which will be the groundwork for all the geometry which is to be created. It is advised to adapt the grid size to the dimensions of the model (e.g., a very small object such as a single fiber will be more visible and easier to operate with when a grid size is also small, e.g., 1 instead of 200) (Figure 3.4).

In order to create a geometry, one uses the graphical tools, which are usually on the left-hand side. This set of tools allows creation, modification, and dimensioning of the geometry. One may create an isolated point, a set of connected lines, circles, rectangles, ellipses, arcs, fillets, splines, offsets curves, or construct horizontal or vertical lines guided through points. In addition, the software allows the setting of constraints to the geometry (e.g., equal lengths of lines or equal angles of the sketch). Other typical graphical tools are also present in this tool bar (e.g., undo last action, redo last action, drag entities, or delete). The software allows the user to add a different sketch and save it.

Each time one points at a selected icon with the cursor, the name of the tool appears in the yellow frame beside the selected tool (e.g., reset view tool) (Figure 3.5) which magnifies the sketch.

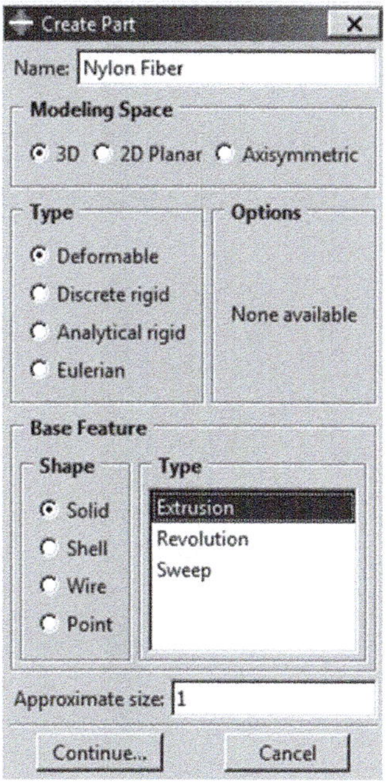

FIGURE 3.4
Create Part submenu allowing the creation of a geometry. © Dassault Systèmes, a French "société européenne" (Versailles Commercial Register # B 322 306 440), or its subsidiaries in the U.S. and/or other countries.

- For simplicity of the modeling process, one assumes a circular cross section for the nylon fiber. Thus, a specific tool *Create Circle* is selected—see Figure 3.6.

Having clicked on *Create Circle,* one can draw the circle and fix its radius. Click on the grid in the central part of the screen to fix an arbitrary center for the circle. This launches the circle shape and initiates sketching. After a final click, the circle appears on the grid. The dimension of the circle can be fixed by the selection of a tool called *Add Dimension,* as presented in Figure 3.7.

It is very important to use values with consistent units, as Abaqus CAE does not have a built-in unit verifier or any type of unit converter. Therefore, the software operator needs to assign the units carefully. As it was suggested by the most reliable guide (Getting Started with the Abaqus, 2016), selecting and rigidly following one selected unit system throughout—SI, SI (mm), US unit (foot), and US unit (inch)—is crucial for receiving a reliable outcome. There is no predefined system of units within Abaqus, so it is a user who is responsible for ensuring that the correct values are specified. Being consistent in using the same unit system is crucial.

In order to establish the radius, one clicks on the shape of the circle (Figure 3.8). This opens a new option situated in the prompt area, which is

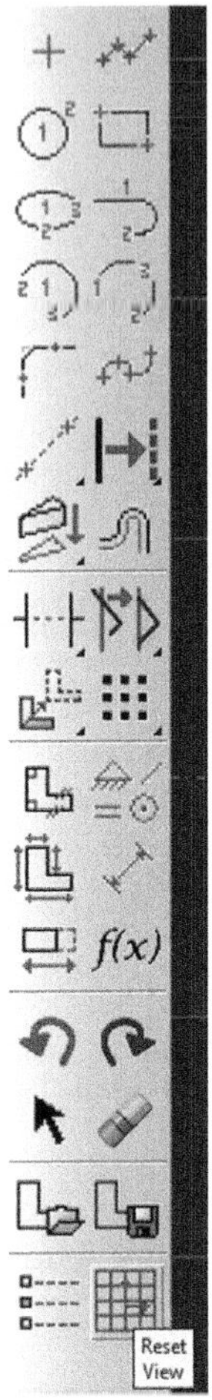

FIGURE 3.5
A set of graphical tools allowing the creation of geometry. © Dassault Systèmes, a French "société européenne" (Versailles Commercial Register # B 322 306 440), or its subsidiaries in the U.S. and/or other countries.

below the grid and above the message area. All comments concerning the current status of processing the model can be found here.

Selecting *Add Dimension* and using the cursor to select the circle launches a new option *New Dimension,* where one may input a radius value (e.g., 0.0005). In this case, one selects the SI unit system, so the radius unit is meter [m]. Next, click the Enter button on your keyboard to confirm the radius value. It will automatically resize the circle on the grid.

The software allows the user to continue dimensioning the geometry by leaving the option *Select the entity to dimension* open under the grid as in Figure 3.9.

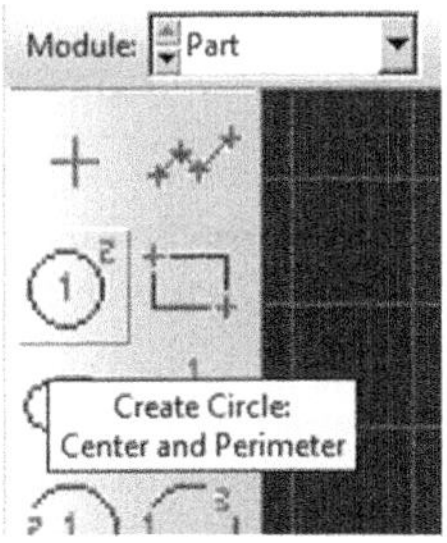

FIGURE 3.6
Selection of the tool *Create Circle* for creation of the geometry. © Dassault Systèmes, a French "société européenne" (Versailles Commercial Register # B 322 306 440), or its subsidiaries in the U.S. and/or other countries.

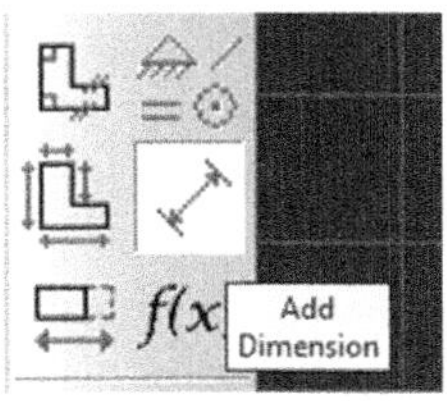

FIGURE 3.7
Selection of a tool called *Add Dimension*, which allows the assignment of desired dimensions to the geometry. © Dassault Systèmes, a French "société européenne" (Versailles Commercial Register # B 322 306 440), or its subsidiaries in the U.S. and/or other countries.

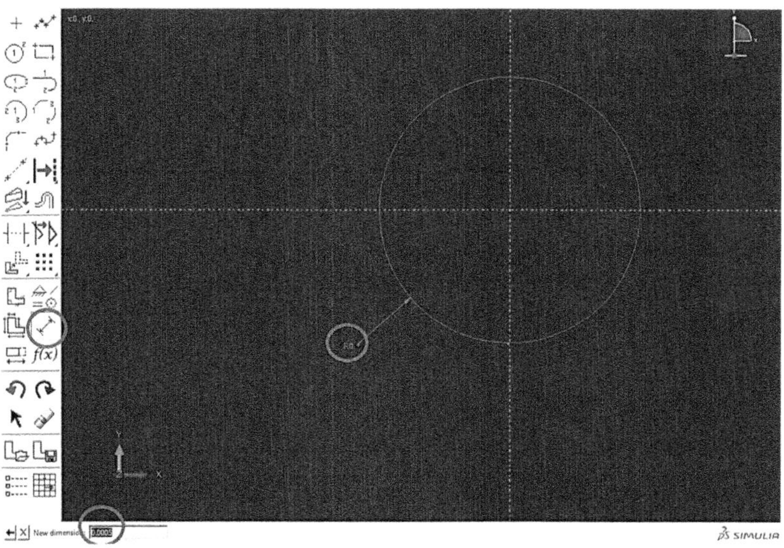

FIGURE 3.8
Sketching the circle and assigning a dimension to it (in this case, a radius). © Dassault Systèmes, a French "société européenne" (Versailles Commercial Register # B 322 306 440), or its subsidiaries in the U.S. and/or other countries.

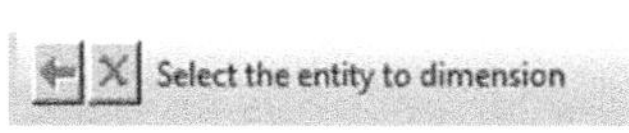

FIGURE 3.9
A tip presented in the prompt area that indicates the possibility of selecting an entity to dimension. © Dassault Systèmes, a French "société européenne" (Versailles Commercial Register # B 322 306 440), or its subsidiaries in the U.S. and/or other countries.

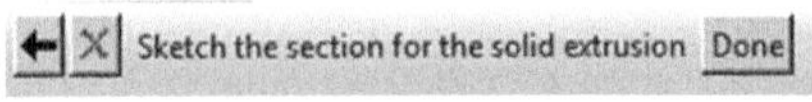

FIGURE 3.10
A tip presented in the prompt area that indicates the need to perform an extrusion step by pressing *Done*. © Dassault Systèmes, a French "société européenne" (Versailles Commercial Register # B 322 306 440), or its subsidiaries in the U.S. and/or other countries.

Clicking on the red cross indicates that one does not wish to introduce any further dimension changes at this point and evokes a new step, which is creation of solid body by extrusion (Figure 3.10).

In order to create a solid body to be the model of a nylon fiber, one should select *Done*. This opens a new window called *Edit Base Extrusion*, which allows selection of the depth of the extrusion. In reality, one selects the length of nylon fiber. A 1 cm length (equal to 0.01 m) is suggested. Other values may be input if there is a need to model longer elements. However, it might be difficult to operate with a fiber dimension model, where the diameter is very small compared to its length (Figure 3.11).

After selecting the OK, a model of the nylon fiber appears on the screen as shown in Figure 3.12.

In order to achieve optimum visibility of the model, depending on the personal preferences, one can use the tools presented in Figure 3.13.

STEP 2: Assigning material properties to the geometry

In order to assign material to the created geometry, one needs to select the material editor's menu bar, which is situated just under *Part*, in the menu tree on the left-hand side of the screen (selection by double-click). As soon as the *Materials* option has been selected, a new window appears on the screen (Figure 3.14).

The user can define not only the materials to be used for the model, but by selecting only the *Mechanical* or only the *Thermal* submenu, one can predefine the type of analysis to be performed. In the case of the current model, the task is to define some of the tensile parameters of the fiber. Therefore, one selects the *Mechanical* submenu. A nylon fiber can be treated as an elastic solid body. Therefore, one should select the *Elasticity* and *Elastic* submenus, remembering

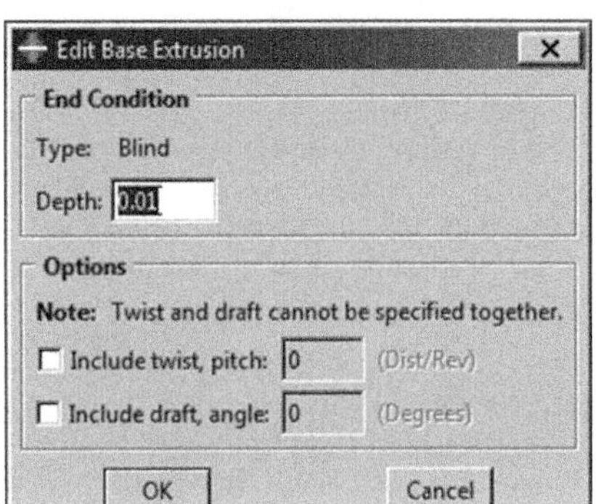

FIGURE 3.11
Edit Base Extrusion submenu. © Dassault Systèmes, a French "société européenne" (Versailles Commercial Register # B 322 306 440), or its subsidiaries in the U.S. and/or other countries.

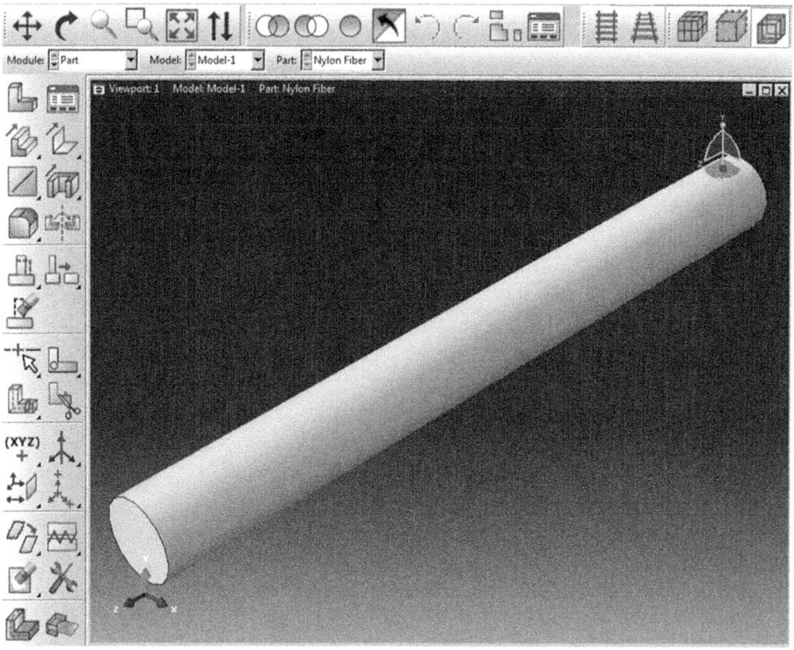

FIGURE 3.12
A viewport from Abaqus/CAE 6.13-3 with a model of a nylon fiber. © Dassault Systèmes, a French "société européenne" (Versailles Commercial Register # B 322 306 440), or its subsidiaries in the U.S. and/or other countries.

FIGURE 3.13
A set of selected tools allowing the user to change the view of the model (left to right: pan view, rotate view, magnify view, and box zoom view). © Dassault Systèmes, a French "société européenne" (Versailles Commercial Register # B 322 306 440), or its subsidiaries in the U.S. and/or other countries.

to be consistent in terms of units and values to be used in setting up the model. Getting Started with Abaqus: Interactive Edition (6.13) presents an overview of unit systems that can be used by Abaqus users. We have already used meters [m] when creating the sketch. To remain consistent, we should therefore use the following SI units for this specific modeling process—length [m], force [N], mass [kg], time [s], stress [Pa = N/m^2], energy [J], and density [kg/m^3].

Abaqus requires us to assign a density to the material to be modeled. The density of nylon fiber is about 1,140 kg/m^3 (Kipp, 2010). This can be input as shown in Figure 3.14.

The next submenu requires Young's modulus and Poisson's ratio for nylon. In this case, the value of Young's modulus that needs to be typed in is 2.5 GPa,

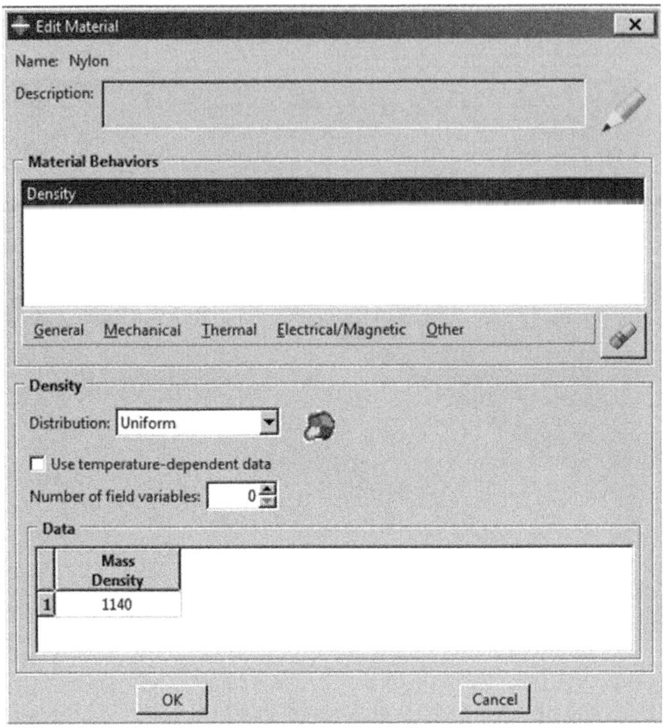

FIGURE 3.14
Edit Material submenu with *Mass Density* of 1140 [kg/m^3] input as the density of the modeled fiber.

and the value of Poisson's ratio is 0.34. Abaqus/CAE displays the elastic data form as shown in Figure 3.15.

As we know, for most materials, a rod-like specimen subjected to uniaxial tension will exhibit some shrinkage in the lateral (radial) direction and elongation in the direction, in which the uniaxial tension is applied. As outlined earlier, the ratio of lateral strain to axial strain is defined as Poisson's ratio. Poisson's ratio for most metals falls between 0.25 and 0.35. Rubber has a Poisson's ratio of 0.499 and is therefore almost incompressible. Theoretical materials with a Poisson's ratio close to 0.5 are truly incompressible, since the sum of all their strains leads to a zero volume change (Mott and Roland, 2009). If Poisson's ratio is close to 0.5, which is the case of rubber-like materials, the bulk modulus (substance's resistance to uniform compression) exceeds the shear modulus (ratio of shear stress to the shear strain) and the material is considered as incompressible. If Poisson's ratio is close to –1, the material becomes highly compressible, and its bulk modulus is less than its shear modulus (Laker, 1987). Thus, Poisson's ratio is bounded by two theoretical limits: it must be greater than –1 and less than or equal to 0.5. If, for some

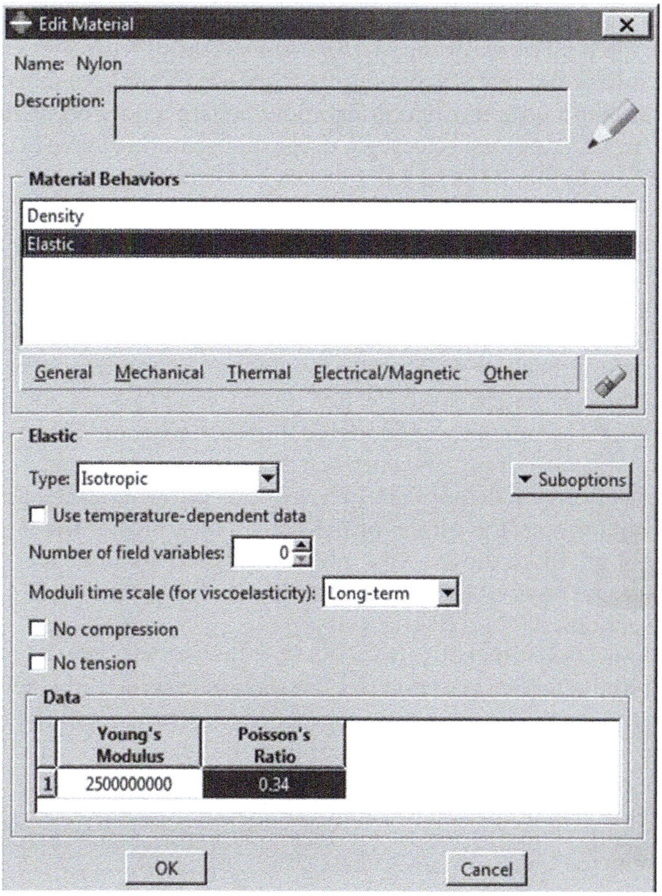

FIGURE 3.15
Edit Material submenu with a Young's modulus of 2.5 GPa and a Poisson's ratio of 0.34, which characterize nylon (polyamide). © Dassault Systèmes, a French "société européenne" (Versailles Commercial Register # B 322 306 440), or its subsidiaries in the U.S. and/or other countries.

reason, one inputs an invalid Poisson's ratio (e.g., 0.51), Abaqus CAE reports an error, and its feedback will be "An invalid Poisson's ratio value has been specified. The Poisson's ratio must be less than the value 0.5."

Click OK to exit the material editor and to finish this step.

The next step is Section creation, which is situated below the Calibration module in the *Model tree* on the left side of the screen.

STEP 3: Section

In the *Model tree*, double-click *Sections* to create a section, which combines the type of material and its properties with the created geometry. A small dialog window called *Create Section* appears. It contains several choices.

For a nylon fiber, the best choice is a *Solid* and *Homogenous* type of modeled object. We will call our section as *Fiber section* (Figure 3.16).

After confirming our selections by clicking *Continue* a new submenu appears. This time, it is a small window where one selects the type of material—in this case, we select nylon. If other materials have been defined, the arrow next to the Material text box can be clicked to see a list of available materials and to select the material of one's choice. However, as *Nylon* is the only material-related file created previously, it appears automatically. Click OK for this selection. In this step, one associates the created geometry with the type of model and material properties (Figure 3.17).

The next step, which is assigning the section to the solid body one created, requires coming back to *Parts* in the *Model tree*, expanding it, and double-clicking *Section Assignment*.

In the prompt area below the model of the fiber, we find "Select the regions to be assigned a section." The software requires selection of the region to which the section will be applied. In case of the current model, we select the whole by clicking on the model of the fiber. The edges of the whole model will be highlighted. In the case of a model with many regions that are to have the same section assigned, one may create a set of regions, which can all have the same section assigned (Figure 3.18).

By clicking *Done*, one confirms the selection and creation of a single set containing the whole region that has the section assigned to it. One can toggle off the option *Create set* in the prompt area, so the set will not be created.

Another small submenu window appears. It is the *Edit Section Assignment* dialog box containing existing sections (Figure 3.19).

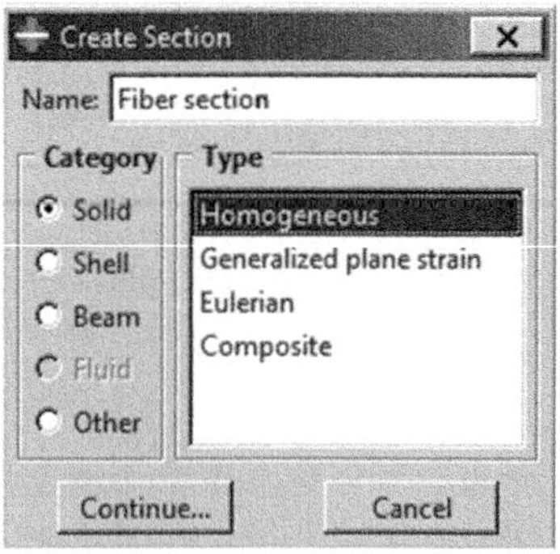

FIGURE 3.16
Create Section submenu. © Dassault Systèmes, a French "société européenne" (Versailles Commercial Register # B 322 306 440), or its subsidiaries in the U.S. and/or other countries.

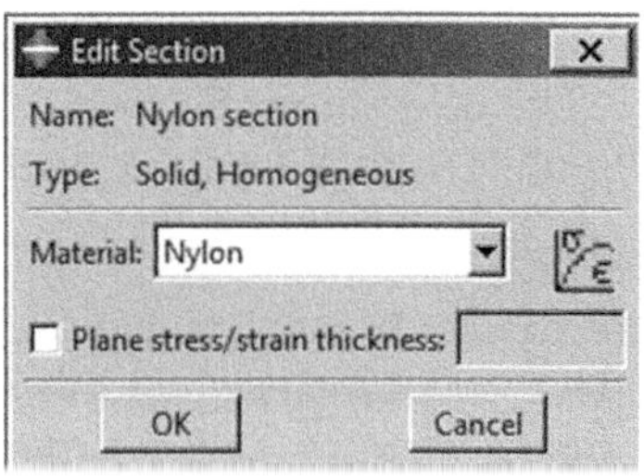

FIGURE 3.17
Edit Section submenu. © Dassault Systèmes, a French "société européenne" (Versailles Commercial Register # B 322 306 440), or its subsidiaries in the U.S. and/or other countries.

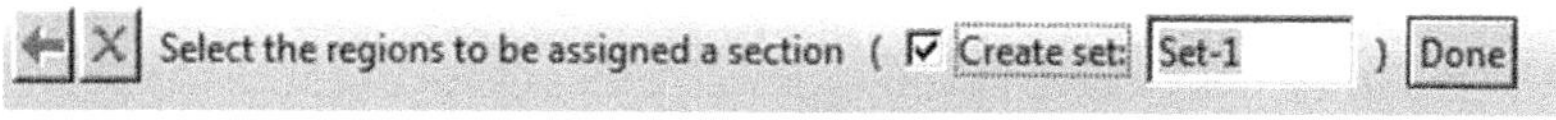

FIGURE 3.18
Prompt area comment requiring the selection of regions to have the section assigned to them. © Dassault Systèmes, a French "société européenne" (Versailles Commercial Register # B 322 306 440), or its subsidiaries in the U.S. and/or other countries.

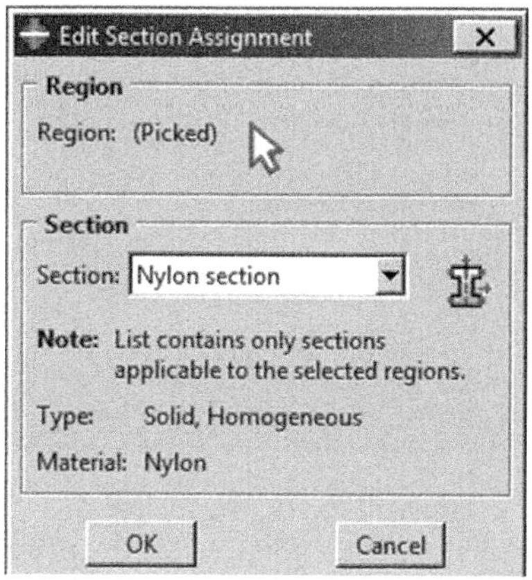

FIGURE 3.19
Edit Section Assignment submenu. © Dassault Systèmes, a French "société européenne" (Versailles Commercial Register # B 322 306 440), or its subsidiaries in the U.S. and/or other countries.

This also acts as a preview of the last few steps taken. The software also presents a comment in the prompt area requiring the filling out of the *Edit Section Assignment* dialog window (Figure 3.20).

By clicking on OK in the *Edit Section Assignment* submenu window, one confirms the selection. Abaqus assigns the section to the solid body frame, colors the entire frame light green to indicate that the region has a section assignment, and closes the Edit Section Assignment dialog box.

STEP 4: Defining the assembly

Each part that you create is oriented in its own coordinate system and is independent of the other parts of the model. Although a model may contain many parts, it contains only one assembly. One defines the geometry of the assembly by creating instances of a part and then positioning the instances relative to each other in a global coordinate system. An instance

FIGURE 3.20
Prompt area comment requiring the filling out the *Edit Section Assignment*. © Dassault Systèmes, a French "société européenne" (Versailles Commercial Register # B 322 306 440), or its subsidiaries in the U.S. and/or other countries.

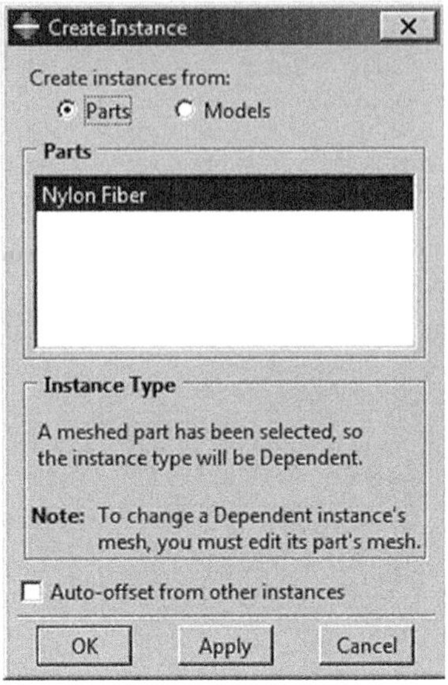

FIGURE 3.21
Create Instance submenu. © Dassault Systèmes, a French "société européenne" (Versailles Commercial Register # B 322 306 440), or its subsidiaries in the U.S. and/or other countries.

can be classified as either independent or dependent (of the original part). Independent part instances are meshed individually while the mesh of a dependent part instance is associated with the mesh of the original part. For further details, see "Working with part instances" of the Abaqus/CAE User's Guide.

Double-click on Instances in Assembly in the *model tree*. Abaqus CAE automatically switches to the *Assembly*, and the *Create Instance* submenu appears as shown in Figure 3.21.

Accept the default proposition as the only part present in the viewport is a *nylon Fiber* model. Its color has changed to blue, and its edges are highlighted in red. The instance is created based on the *Parts* element. Click OK to confirm the selection.

STEP 5: Step creation

This model does not "know" what kind of action will be required to perform (e.g., whether it is going to be a thermal simulation or mechanical one). In order to establish this, it is necessary to double-click on the "Steps" mode in the *Model tree* on the left side of the screen.

A new dialog window appears, and a new action is required. We need to give it a title (e.g., *Fiber deformation*). Having given a name to the new step (note that the *Initial* step is created automatically by the software), select a procedure type: *General* and *Static, General* and click on *Continue* to open another dialog window. In this new submenu, one may propose a different description than *Deformation*. Figure 3.22a and b shows the creation and editing of the step called *Fiber deformation*.

It is advised that the relation between the introduced time frame and the number of increments and the output of the model are observed. In *Incrementation*, one needs to make sure that the required data are provided: *Maximum number of increments* and *Increment size* values. See Figure 3.23.

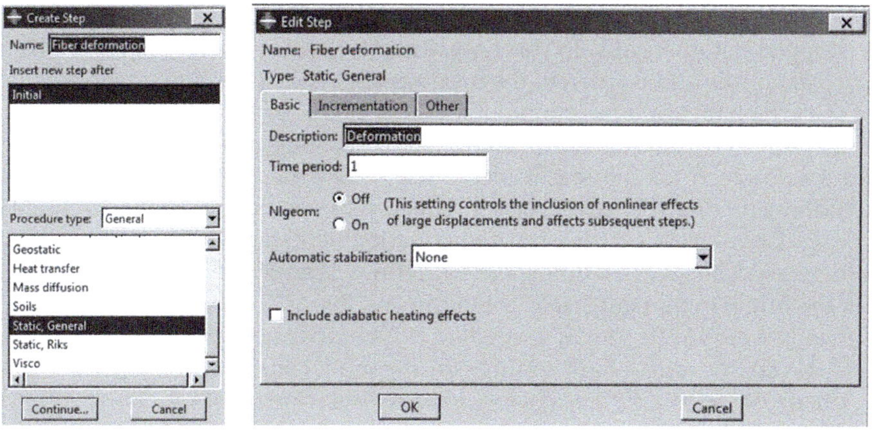

(a) (b)

FIGURE 3.22
(a) *Create Step* submenu and (b) *Edit Step* submenu. © Dassault Systèmes, a French "société européenne" (Versailles Commercial Register # B 322 306 440), or its subsidiaries in the U.S. and/or other countries.

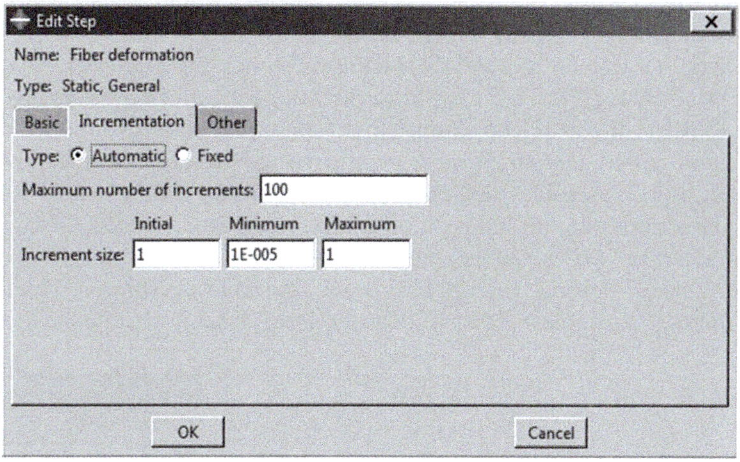

FIGURE 3.23
Edit Step submenu. © Dassault Systèmes, a French "société européenne" (Versailles Commercial Register # B 322 306 440), or its subsidiaries in the U.S. and/or other countries.

Abaqus provides with an *Automatic* assessment of values which may be changed or customized. The beginner is advised to accept the automatic time incrementation and experiment by modifying it later on to learn how changing the incrementation influences the results. By default, Abaqus will set the maximum number of increments to 100. The *Maximum number of increments* is the largest number of small steps required by the equations governing the model to complete the calculation for the given step, which we have called *Fiber deformation*. If the step exceeds this number of increments, calculation is stopped automatically. In this case, the information about it is reported in the *Job* module and written to the message file. This is situated at the bottom of the screen, below a prompt area, in the message area.

If one selects *Automatic* time incrementation, Abaqus starts the step using the value given for the initial increment size, which, in case of the dialog window *Edit Step* (Figure 3.23), is 1. The size of subsequent time increments is calculated based on how fast the equations for the model can be solved. When one chooses *Fixed* time incrementation, Abaqus uses the value entered for the initial increment size throughout the step.

The last tab in the *Edit Step* is *Other*, which does not require any changes and deviation from default settings for the model currently described.

Click OK so the Edit Step dialog window will close.

STEP 6: Field output requests

Although this step can be completed later, this is a good point at which to select the parameters of the model that one would like to verify, observe, and compare when the model is processed. These are the outcome parameters. The selection of outcome parameters can be performed in *Field Output Requests*. This element of the *Model tree* can be found on the left side of the screen in the tree, after *Steps*. Double-click on *Field Output Requests* and a new dialog window will appear (Figure 3.24).

The default name F-Output-1 can be kept. By clicking on *Continue*, a default dialog window for *Edit Filed Output Request* appears, and we have the opportunity to select the required parameters. There is a large group of different parameters called *Output Variables* comprising Stresses (14 different variables), Strains (21 variables), Displacement/Velocity/Acceleration (8 variables), Forces/Reactions (21 variables), Contact (9 variables), Energy (3 variables), Failure/Fracture (32 variables), Thermal (6 variables), Electrical/Magnetic (12 variables), Porous media/Fluids (6 variables), Volume/Thickness/Coordinates (7 variables), Error Indicators (1 variable), and State/Field/User/

FIGURE 3.24
Create Field submenu. © Dassault Systèmes, a French "société européenne" (Versailles Commercial Register # B 322 306 440), or its subsidiaries in the U.S. and/or other countries.

Time (6 variables). Any of these can be selected and therefore calculated for a modeled nylon fiber.

Here one refers back to the previously created *Step* called *Fiber deformation* and selected procedure: *Static, General*.

As shown in Figure 3.25, one may select a domain that refers to receiving the *Output Variables* for a Whole model or only a selected part. Domain allows selection from: Whole model, Set, Bolt load, Composite layup, Fasteners, Assembled fasteners set, Substructure, Interaction, Skin, and Stringer. In case of this study, one should select *Whole model*. If under Domain, one selects Set, Abaqus automatically provides a list of all the sets created for a given model. However, one did not create any sets for this model.

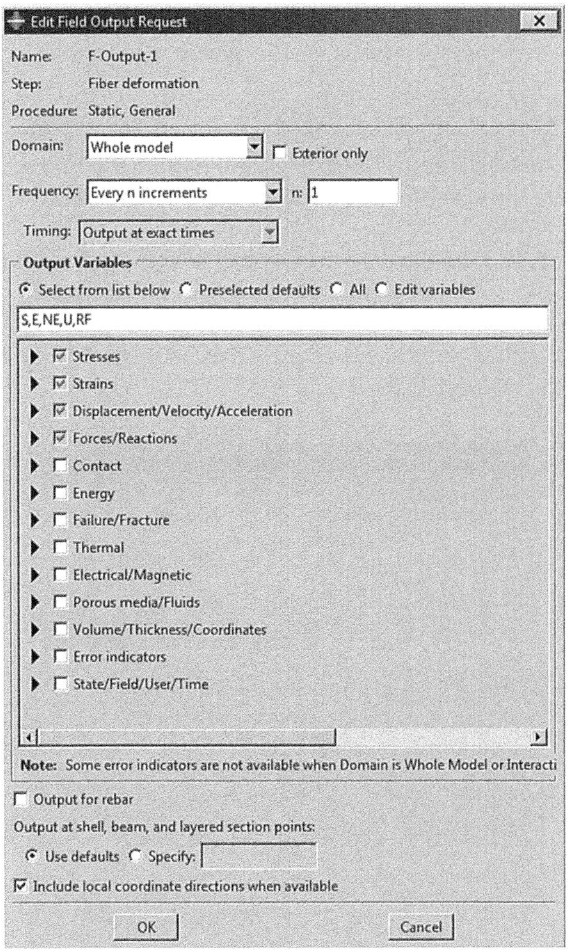

FIGURE 3.25
Edit Field Output Request submenu. © Dassault Systèmes, a French "société européenne" (Versailles Commercial Register # B 322 306 440), or its subsidiaries in the U.S. and/or other countries.

Under the Frequency field, select every *n* time increments and set the value of *n* to 1 to write the output at every increment. Thus, the full calculation for all the increments can be observed.

The Output Variables and the symbols used in the software to identify the variables may be rendered visible by clicking on the black triangular pointer on the left-hand side of the columns with variables. For example, opening the Stresses option allows selection of: Stress components and invariants, S; Mises equivalent stress, MISES; Maximum Mises equivalent stress, MISESMAX; Transverse shear stress (for thick shells), TSHR; Transverse shear in stacked continuum shells, CTSHR, etc.

One proposes the selection of—Stress components and invariants (S), Total strain components (E), Nominal strain components (NE), Translations and rotations (U), and Reaction forces (RT).

When one selects a desired variable, this selection is visible in dialog window.

STEP 7: BCs and Loads

- BCs are the constraints, the existence of which are known, and which need to be imposed to carry out the modeling correctly.

Double-click on BCs in the *Model tree* on the left-hand side of the window. This opens a new submenu window as shown in Figure 3.26.

To simulate a real tensile test, one needs to virtually clamp one of the ends of the nylon fiber and pull the other end. This is the first of the two additional BCs and is called *Pinned end*.

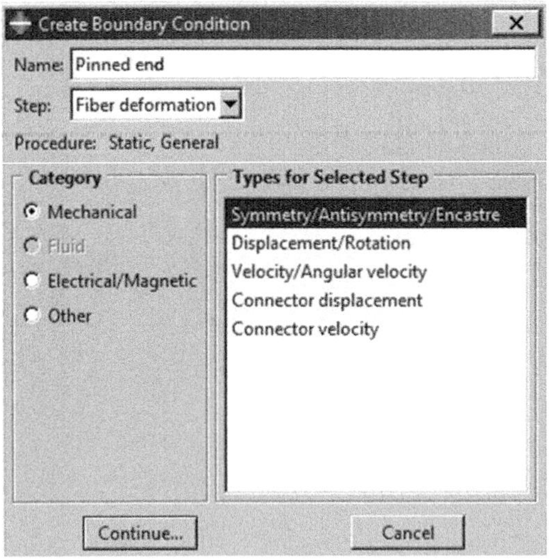

FIGURE 3.26
Create Boundary Condition submenu. © Dassault Systèmes, a French "société européenne" (Versailles Commercial Register # B 322 306 440), or its subsidiaries in the U.S. and/or other countries.

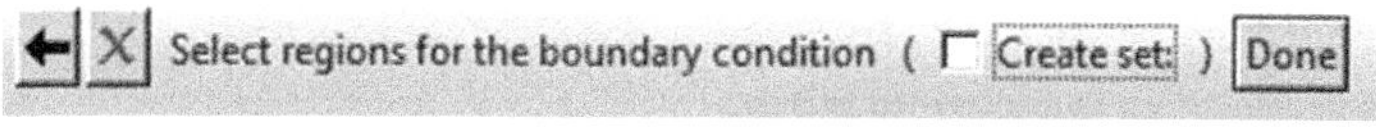

FIGURE 3.27
A prompt requiring the selection of a region for a BC application. © Dassault Systèmes, a French "société européenne" (Versailles Commercial Register # B 322 306 440), or its subsidiaries in the U.S. and/or other countries.

In the submenu window presented in Figure 3.26, one needs to select the step called *Fiber deformation* and the type of step called *Symmetry/Antisymmetry/Encastre*. Selection of this specific type guarantees fixing on the ends of the cylinder.

After clicking on *Continue*, a prompt area presents a comment (Figure 3.27) that requires selection of the region where BC is to be applied.

At this stage, it does not matter which of the two faces is marked as the pinned one, as the object is symmetrical. The selected face is highlighted as presented in Figure 3.28a. Confirm the selection by clicking on *Done* in the prompt area. This action prompts the opening of another submenu (Figure 3.28b), where one edits the type of BC. As was just discussed, the most suitable option is *Pinned*.

In the case of this specific model conditions, one may achieve a similar result by imposing *Pinned* or a full *Encastre*; however, these two BCs do not mean the same and should not be applied interchangeably as they reflect different BCs that are presented in Figure 3.28b.

Clicking on OK confirms the selection. Two marks indicating imposed pinned BCs are seen in Figure 3.28c.

This is a convenient moment to explain the meaning of terminology related to BCs.

Some BCs impose restrictions to the model eliminating any displacement (U) and/or displacement and rotation (UR), which means that U and/or UR equal to zero (e.g., U1 = U2 = U3 = 0 and U1 = U2 = U3 = UR1 = UR2 = UR3 = 0).

It is also suggested that all other elements of the model without a specified BC have nonzero displacements and/or rotations. Abaqus CAE uses digits 1, 2, and 3 to refer to the specific axes, along which the movement of the model is possible or it is not possible because it was restricted. Figure 3.29 presents the configuration of axes on the background of the model.

Any displacement in the direction of axis x is termed U1, and any rotation about this axis is termed UR1.

Any displacement in direction of axis y is termed U2, and any rotation about this axis is termed UR2.

Any displacement in direction of axis z is termed U3, and any rotation about this axis is termed UR3.

The second BC can be imposed on the model after double-clicking on BCs in the *Model tree* on the left-hand side of the window. This action opens the submenu window presented in Figure 3.26. This time, one names it *Unidirectional movement* and selects *Displacement/Rotation* type of Step, instead of *Symmetry/Antisymmetry/Encastre*. Leaving U3 parameter

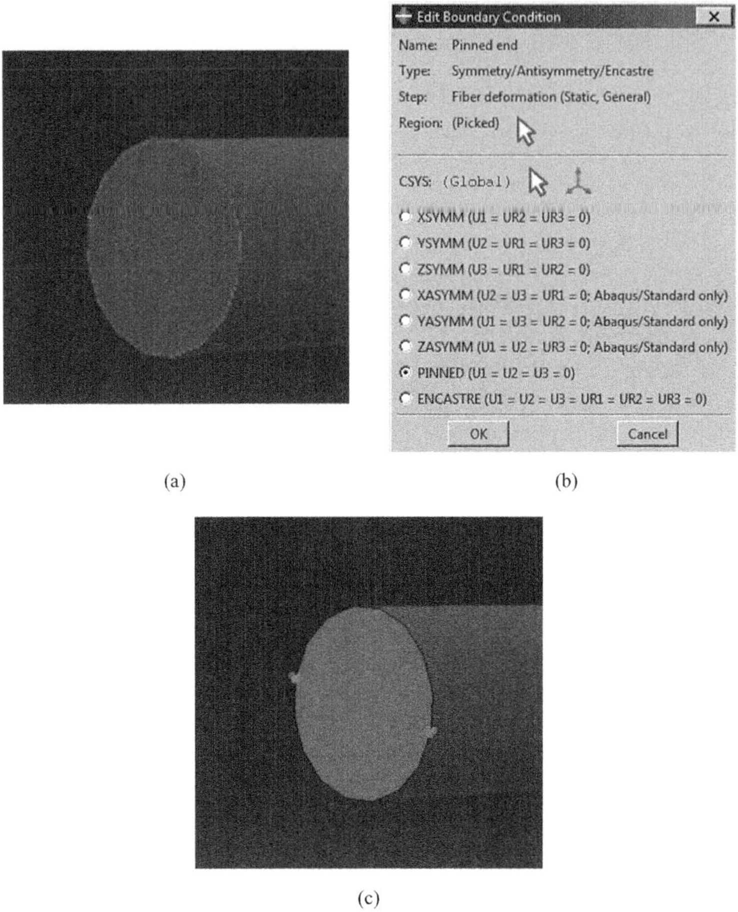

FIGURE 3.28
(a) Highlighted edge of the cylinder that represents imposing *Pinned* as a BC, (b) *Edit Boundary Condition selection* submenu—*Pinned,* and (c) graphical presentation of imposed BC Pinned to one face of the cylinder. © Dassault Systèmes, a French "société européenne" (Versailles Commercial Register # B 322 306 440), or its subsidiaries in the U.S. and/or other countries.

without any restrictions allows movement of the the second edge of the cylinder. After clicking on *Continue,* one needs to select the region where BC is to be applied. At the same time, a prompt area presents the comment (Figure 3.27) that requires selection of the region where BC is to be applied. This time one is asked to select the other face as the previous face is already *Pinned*. This action prompts the opening of another submenu as presented in Figure 3.30, where one edits the type of BC.

The simulation of a tensile test performed on the nylon fiber requires the simulation of two clamps. The first one, which is fixed, is represented by a *Pinned end*. The second one, the one that pulls the fiber down (to simulate

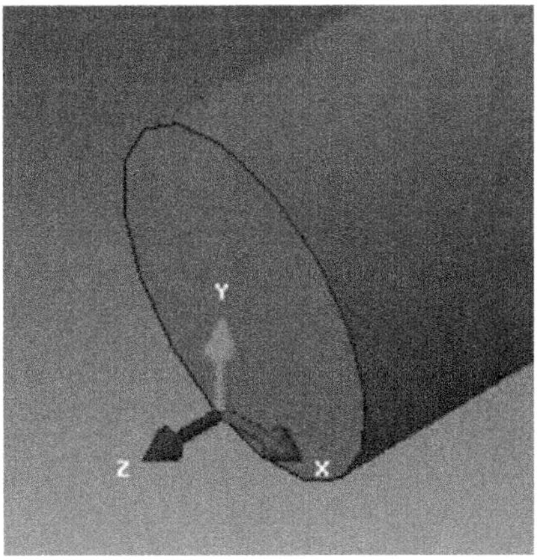

FIGURE 3.29
Axes configuration according to Abaqus CAE. © Dassault Systèmes, a French "société européenne" (Versailles Commercial Register # B 322 306 440), or its subsidiaries in the U.S. and/or other countries.

a tension test) is reflected by *Unidirectional movement* and operates in one direction only, i.e. along z-axis.

Confirming the selection by selecting OK imposes the BC on the other face of the nylon fiber model, which is presented in Figure 3.31.

- Loads

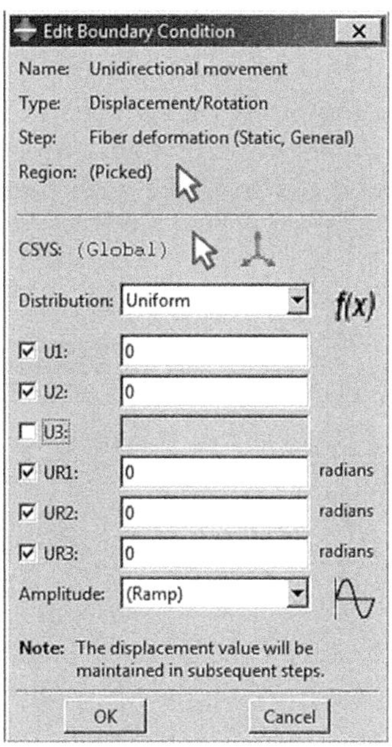

FIGURE 3.30
Edit Boundary Condition submenu—Displacement/Rotation. © Dassault Systèmes, a French "société européenne" (Versailles Commercial Register # B 322 306 440), or its subsidiaries in the U.S. and/or other countries.

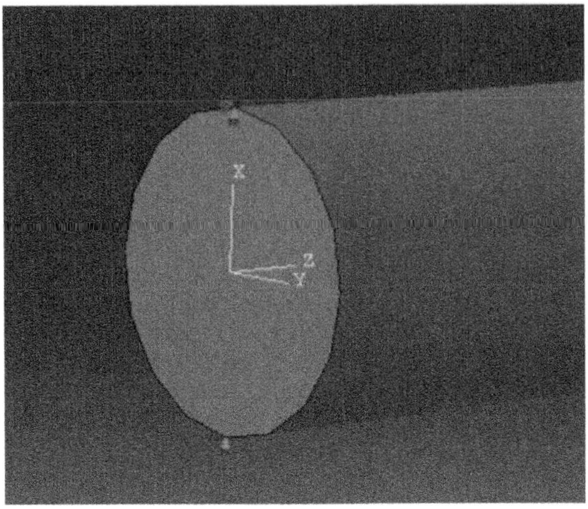

FIGURE 3.31
Graphical presentation of Displacement BC imposed on the second face of the cylinder. © Dassault Systèmes, a French "société européenne" (Versailles Commercial Register # B 322 306 440), or its subsidiaries in the U.S. and/or other countries.

The next part in this step consists in applying a Load to the face, which is to be displaced or pulled down.

Double-click on Load in the *Model tree* on the left side of the window. It opens a new submenu window as presented in Figure 3.32. This Load is given the name Tension. A Step called Fiber deformation should be selected automatically; if not, one needs to make sure that Load is applied into a specific step.

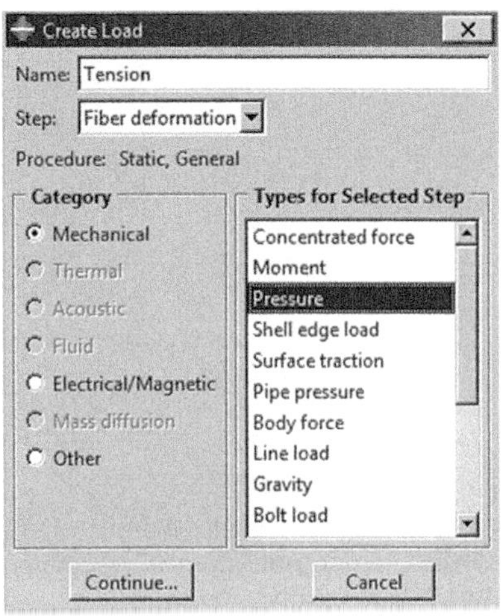

FIGURE 3.32
Create Load submenu. © Dassault Systèmes, a French "société européenne" (Versailles Commercial Register # B 322 306 440), or its subsidiaries in the U.S. and/or other countries.

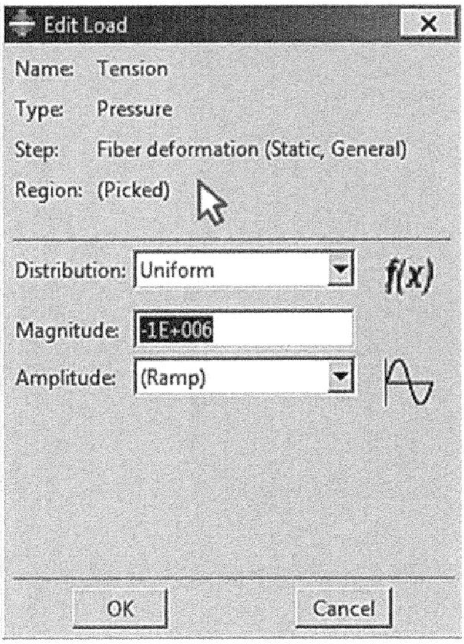

FIGURE 3.33
Edit Load submenu. © Dassault Systèmes, a French "société européenne" (Versailles Commercial Register # B 322 306 440), or its subsidiaries in the U.S. and/or other countries.

Select *Continue* in order to mark the individual element of the model, where the pressure is to be applied. An adequate information appears in the prompt area. One may toggle off the option, allowing for creation of a set of surfaces. As soon as the selection is made, the selected element changes the color into red. Select *Done* in the prompt area to confirm selection and finalize this step. Finalizing this step evokes opening a new submenu window called *Edit Load.*

One selects Mechanical in the category of load and Pressure in the Type for Selected Step.

Among all presented types, Pressure reflects the most correctly the idea of tensile test. If one understands the idea of pressure literally, it is applying a force on a specific surface. However, in case of the current model, tensile test requires pulling the surface. To accomplish this goal, instead of applying 1 MPa on the surface, one applies –1 MPa by typing in the value –1E+006 (Figure 3.33), which changes the direction of the force. Figure 3.34 presents the edge of the model of the fiber with a second BC and applied Load.

STEP 8: Mesh

To mesh a model, which means creating a new representation of a surface or a body of the model by means of geometric shapes (in case of 2D models) and polyhedrons (in case of 3D models), one needs to select the size and the type of this representation.

Click on *Parts (1)* in the *Model tree* on a left-hand side of the window. This opens a submenu containing details of the *Nylon Fiber,* at the end of which one finds the *Mesh* option (Figure 3.35). Note that this step could be completed before imposing *BCs* and *Load* to the model.

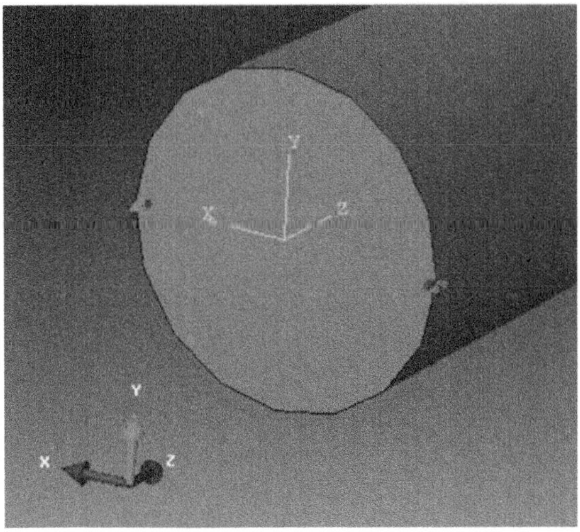

FIGURE 3.34
The edge of the model of the fiber with imposed second BC and applied Load represented by small arrows indicating the direction of force. © Dassault Systèmes, a French "société européenne" (Versailles Commercial Register # B 322 306 440), or its subsidiaries in the U.S. and/or other countries.

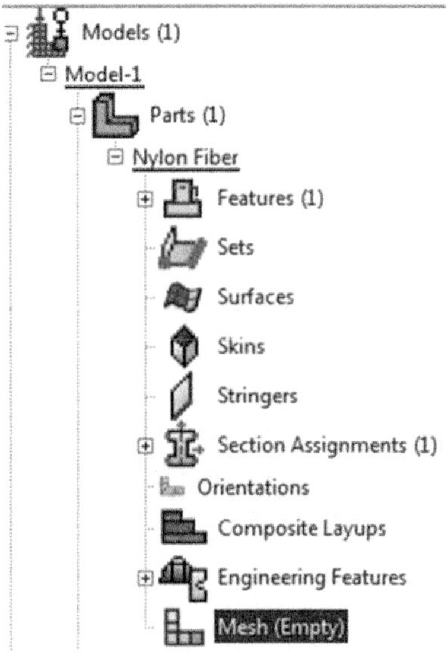

FIGURE 3.35
A fragment of the *Model Tree* showing *Mesh* option. © Dassault Systèmes, a French "société européenne" (Versailles Commercial Register # B 322 306 440), or its subsidiaries in the U.S. and/or other countries.

Double-clicking on Mesh changes the color of the model to yellow. Select *Seed* in the top bar menu or select the appropriate icon, which is usually on the left side of the viewport towards the top. This opens a *Global Seed* submenu window where one can select the size of elements in the mesh. The software automatically adjusts the size of the mesh, which is shown in Figure 3.36.

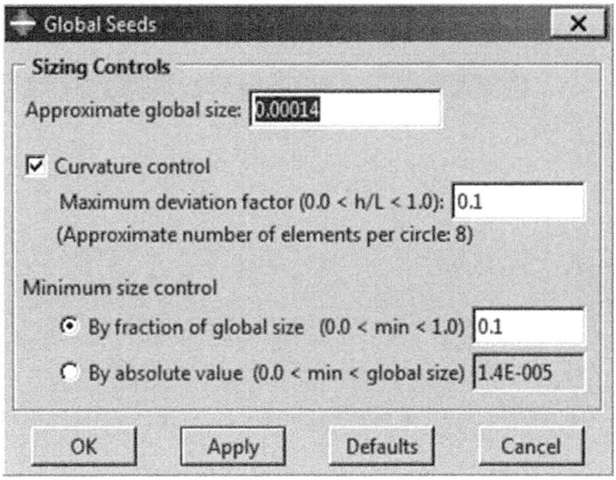

FIGURE 3.36
Global Seeds submenu. © Dassault Systèmes, a French "société européenne" (Versailles Commercial Register # B 322 306 440), or its subsidiaries in the U.S. and/or other countries.

No further adjustment is needed in this particular case. Click OK for confirmation. Next select *Mesh/Element Type* in the top bar menu or select an appropriate icon, which is usually on the left side of the viewport. This opens an *Element Type* submenu window, where one can select the analysis type for this specific type of mesh. In the case of the current model, the 3D Stress option should be selected. Other options in the *Element Type* submenu are shown in Figure 3.37. The shape of the polyhedron can also be selected in the same submenu widow. However in the specific case of the current model, one suggests a *C3D8 (An 8-node linear brick)*, which is one of the most basic options. The suitability of the mesh element can be checked using the option *Verify Mesh* in the *Mesh* menu or selection of the appropriate icon, which is usually on the left side of the viewport.

It helps to verify if the selected mesh is the best option for the meshed geometry and highlights the potentially troublemaking regions where finer or different type of meshing is required.

To complete the *Mesh* step, one needs to select *Mesh/Part* from the top bar menu or select the appropriate icon, which allows meshing a whole model according to selected parameters. In the prompt area, one may find a question: "OK to mesh the part?" Select Yes to confirm meshing, which finalizes this step. The meshed model changes color to light blue. This is shown in Figure 3.38.

In the Message Area one finds the information: *Number of elements: 4331*, which is the number of elements creating the mesh of the model. As none of the specific types of the polyhedron was selected, the mesh contains cubic-like elements as presented in Figure 3.39.

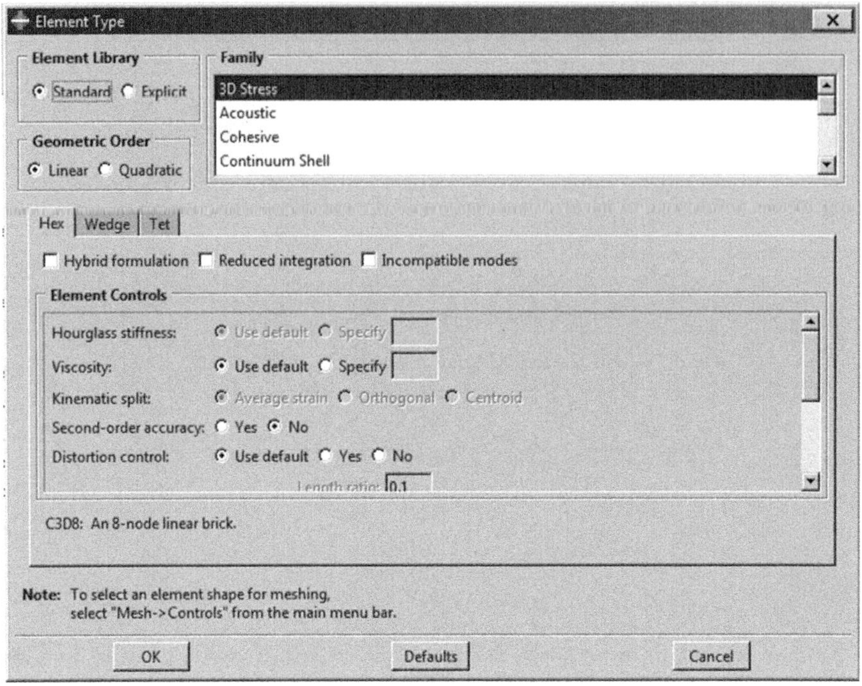

FIGURE 3.37
Element Type submenu. © Dassault Systèmes, a French "société européenne" (Versailles Commercial Register # B 322 306 440), or its subsidiaries in the U.S. and/or other countries.

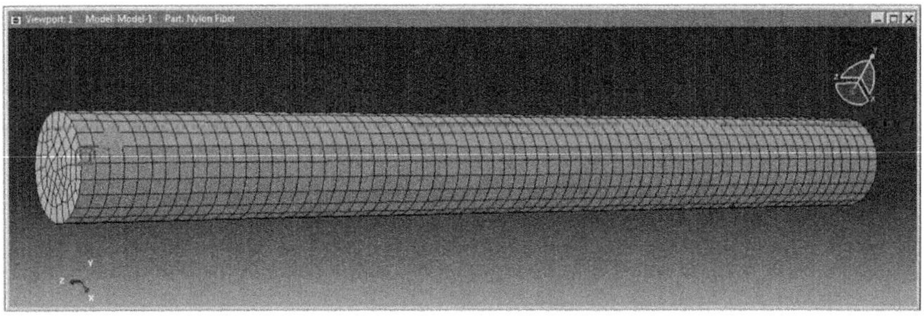

FIGURE 3.38
A meshed model of the nylon fiber. © Dassault Systèmes, a French "société européenne" (Versailles Commercial Register # B 322 306 440), or its subsidiaries in the U.S. and/or other countries.

STEP 9: Job

To verify the model setting and perform the analytical part, which is carried out by Abaqus CAE, one needs to create a *Job*, which is a process of verification and processing of all inputs introduced into the model in the steps from 1 to 8. If any errors are found, an appropriate warning messages are

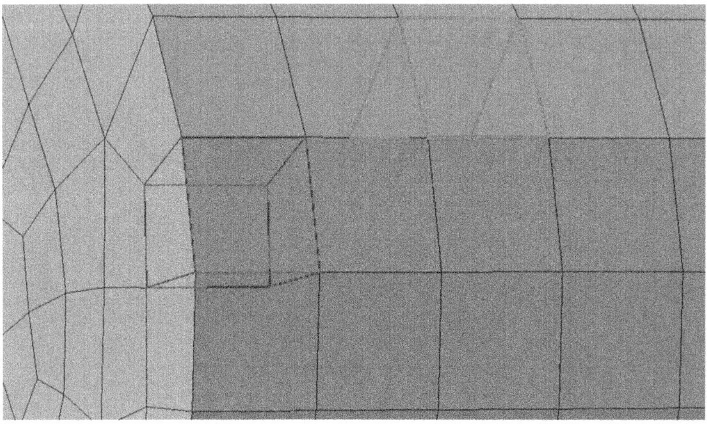

FIGURE 3.39
A magnified fragment of the meshed model of the nylon fiber with two illustrative elements with cubic-like shapes marked. © Dassault Systèmes, a French "société européenne" (Versailles Commercial Register # B 322 306 440), or its subsidiaries in the U.S. and/or other countries.

presented in the message area. The details of the performed analysis of the created model can be found in the Job Monitor dialog box.

In order to create a *Job*, double-click on *Job* in the *Model tree* on the left side of the screen, almost at the end of the tree. A new submenu called *Create Job* appears. One may use the name, *Unidirectional_tension* (no spaces between signs when creating the name), which is proposed in Figure 3.40.

After clicking on *Continue*, another submenu window appears. This time one may edit the job by selecting the name (e.g., *Unidirectional_tension*). At this stage of modeling, and for this specific model, there is no need to use other options: *General*, *Memory*, *Parallelization*, and *Precision*, which exist in this submenu window (Figure 3.41).

Confirm the step by clicking on OK, and select a new submenu by clicking on Unidirectional_tension and selecting *Data Check*, as shown in Figure 3.42. As soon as the data are checked and a confirmation of this step is presented in the Message Area, one can select *Submit*.

As Job processing progresses (the term "running" is used in the *Model tree*), one may observe its status and monitor it by selecting *Monitor*.

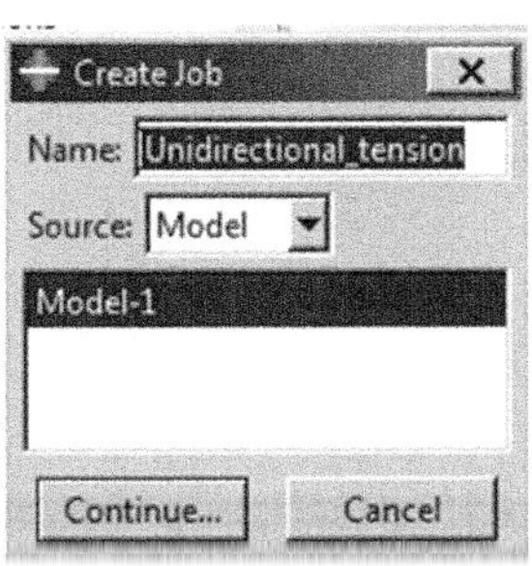

FIGURE 3.40
Create Job submenu. © Dassault Systèmes, a French "société européenne" (Versailles Commercial Register # B 322 306 440), or its subsidiaries in the U.S. and/or other countries.

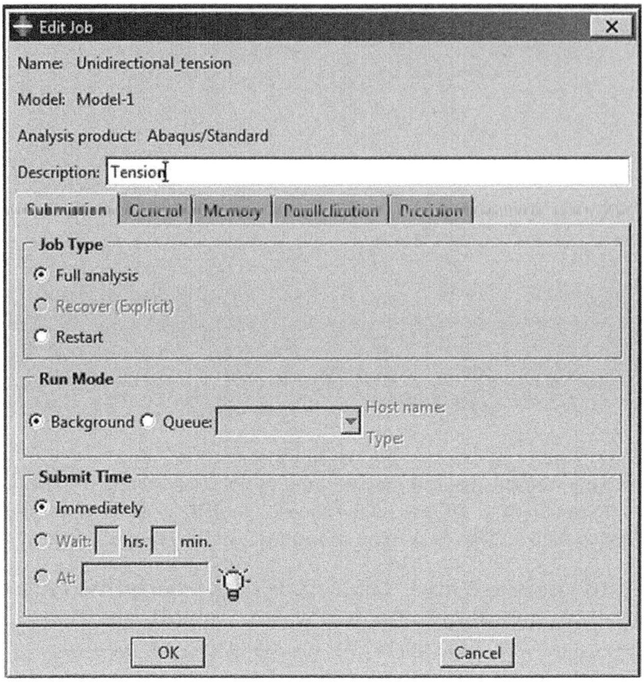

FIGURE 3.41
Edit Job submenu. © Dassault Systèmes, a French "société européenne" (Versailles Commercial Register # B 322 306 440), or its subsidiaries in the U.S. and/or other countries.

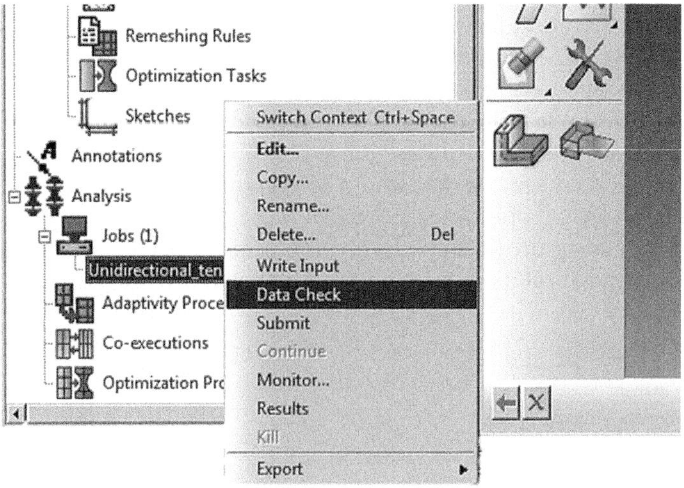

FIGURE 3.42
A fragment of a viewport with a selection of *Data Check*. © Dassault Systèmes, a French "société européenne" (Versailles Commercial Register # B 322 306 440), or its subsidiaries in the U.S. and/or other countries.

When complete the comment in the Message Area says "Job Unidirectional_tension completed successfully." One may now click on *Results*.

STEP 10: Visualization, Results, and Report.

Observation of the results and preparing an appropriate report are important steps at the postmodeling stage as one needs to verify the data, compare it to the mathematical analysis, and find out whether any discrepancies appear between assumptions made at the beginning, any mathematical analysis performed before modeling, and the results presented by Abaqus CAE.

After clicking on *Results*, the software changes the viewport. The new viewport is presented in Figure 3.43. Apart from a meshed, undeformed model of the nylon fiber, which occupies the major part of the screen, one can see two interesting sets of tools. The first, situated on the extreme left-hand

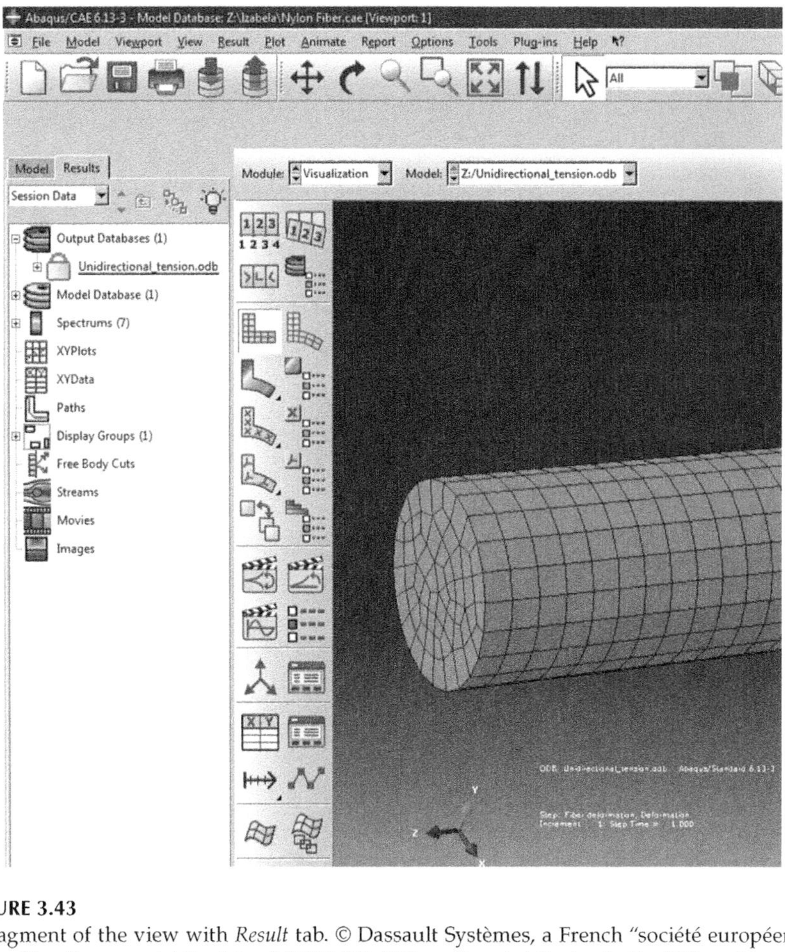

FIGURE 3.43
A fragment of the view with *Result* tab. © Dassault Systèmes, a French "société européenne" (Versailles Commercial Register # B 322 306 440), or its subsidiaries in the U.S. and/or other countries.

side, is actually a tree that has changed its appearance. It is not a *Model tree* any more, but a *Result tree*, with *Output Database* at the pinnacle.

Viewing the results can be done by selecting the icon *Plot Contours on Deformed Shape*, which switches the model into its deformed version.

A field output dialog allows us to select the parameter we wished to observe (e.g., U (deformation)). In other words, one would like to observe how a model of a nylon fiber was deformed after the application of BCs and Load as described in STEP 7: BCs and Loads. The set of figures from 3.44 to 3.49 presents views of a model of a nylon fiber depending on the observed parameters, which were selected in STEP 6: Field Output Requests.

Abaqus CAE presents a legend for each observed parameter and a description under each model. These features can be easily formatted by using an option called *Viewport/Viewport Annotation Options*, which can be found in a tool bar at the top.

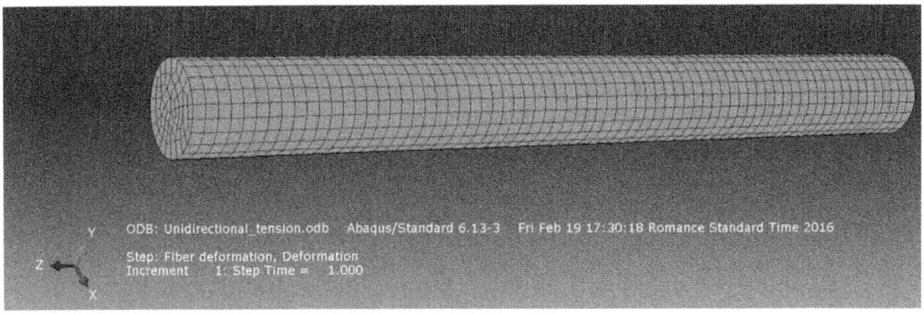

FIGURE 3.44
A model of the undeformed nylon fiber. © Dassault Systèmes, a French "société européenne" (Versailles Commercial Register # B 322 306 440), or its subsidiaries in the U.S. and/or other countries.

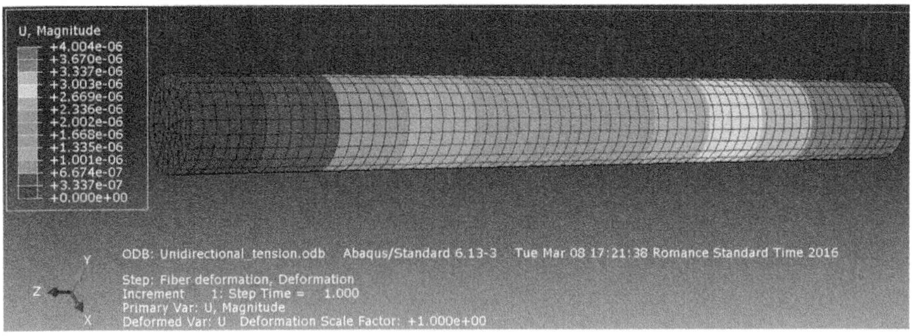

FIGURE 3.45
The magnitude of displacement (U) of one of the faces of the model. The maximum magnitude is marked in red and the minimum is marked in blue. © Dassault Systèmes, a French "société européenne" (Versailles Commercial Register # B 322 306 440), or its subsidiaries in the U.S. and/or other countries.

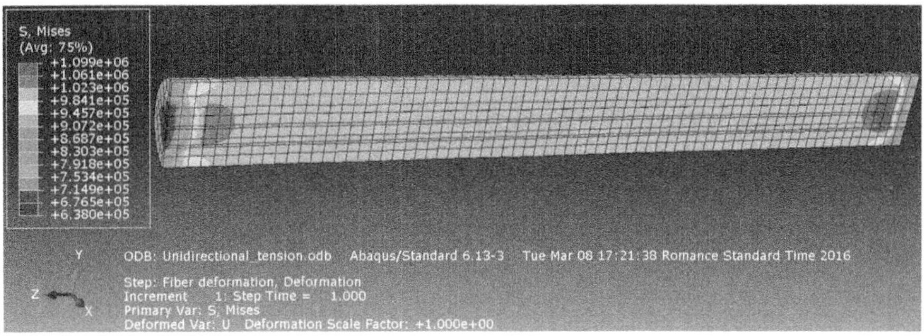

FIGURE 3.46
A longitudinal cross section of the model with Von Mises stress. © Dassault Systèmes, a French "société européenne" (Versailles Commercial Register # B 322 306 440), or its subsidiaries in the U.S. and/or other countries.

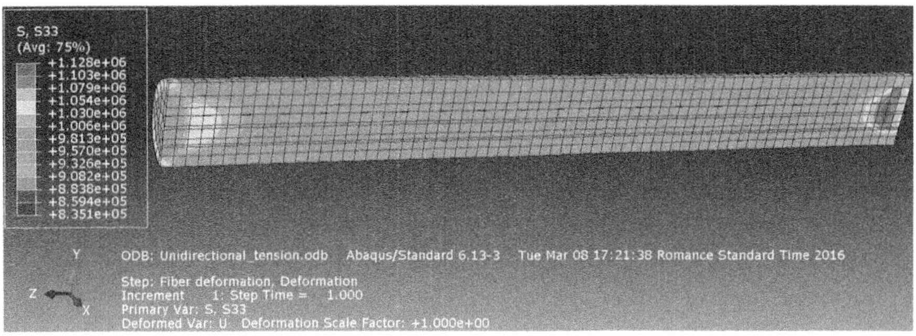

FIGURE 3.47
A longitudinal cross section of the model with stress along *z*-axis. © Dassault Systèmes, a French "société européenne" (Versailles Commercial Register # B 322 306 440), or its subsidiaries in the U.S. and/or other countries.

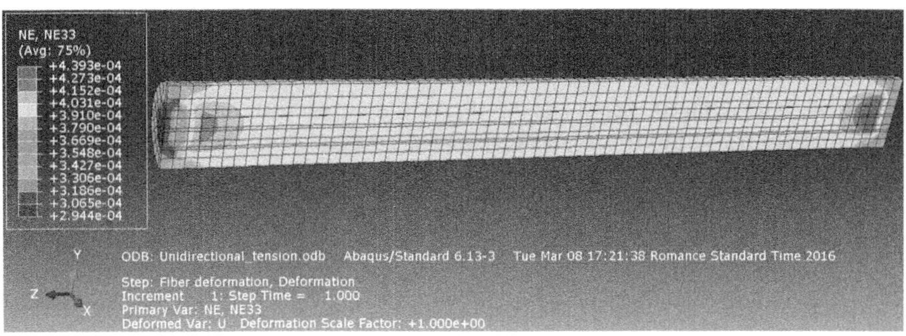

FIGURE 3.48
A longitudinal cross section of the model indicating Nominal Strain (NE) along *z*-axis, which is the ratio of the change in length to the length in the reference configuration in the principal direction, thus giving a direct measure of deformation. © Dassault Systèmes, a French "société européenne" (Versailles Commercial Register # B 322 306 440), or its subsidiaries in the U.S. and/or other countries.

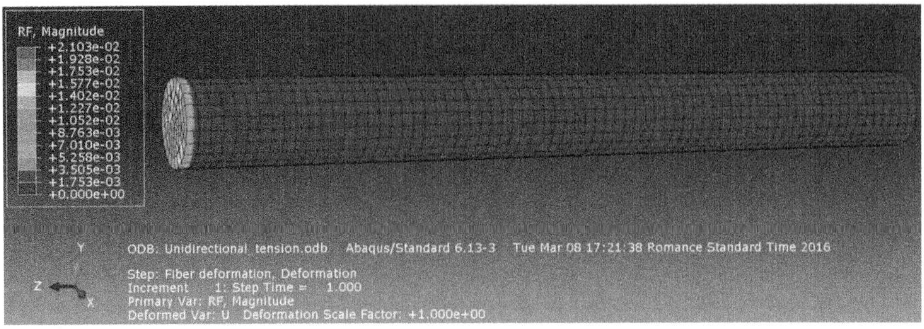

FIGURE 3.49
The magnitude of the reaction force (RF). © Dassault Systèmes, a French "société européenne" (Versailles Commercial Register # B 322 306 440), or its subsidiaries in the U.S. and/or other countries.

As it is depicted in Figure 3.45, the maximum displacement equals 4.004E–6, which is about 0.004 mm, being a calculated displacement value ΔL.The term Von Mises stress is widely used in reference to stress analysis in finite element modeling type of analysis using Abaqus CAE and similar software. A 3D model of a nylon fiber created in Abaqus CAE is an elastic body subject to loads in a 3D Cartesian coordinate system, where a 3D system of stresses and strains is developed. Thus, at any point within the modeled body, stresses may act in different directions, and the direction and magnitude of stresses may change from point to point. The Von Mises stress is utilized to predict yielding of materials under loading condition. In spite of the fact that none of the principal stresses, acting in one direction only (along one of the axes) exceeds the yield stress of the material, it is possible for yielding to result from the combination of stresses. The Von Mises criteria is a formula for combining these three separately acting stresses into an equivalent stress, which is then compared to the yield stress of the material.

This longitudinal cross section can be obtained by selecting the icon called *Active/Deactivate View Cut*.

Reaction force is the force appearing at the constrained end, which should be the absolute value equivalent of the action force imposed to the system, but working in the opposite direction.

- Report

Summary of the analytical part containing changes of selected parameters can be found in the report, preparation of which can be customized. Go to *Report/Field Output* in the main tool bar on the top of the screen (Figure 3.50).

Output variables can be selected from the list of three groups of variables, which are LE, NE, and S. Only these variables are present, because only these parameters were preselected in STEP 6. One may see and select the variables that can be listed in the report by developing a list of options and toggling

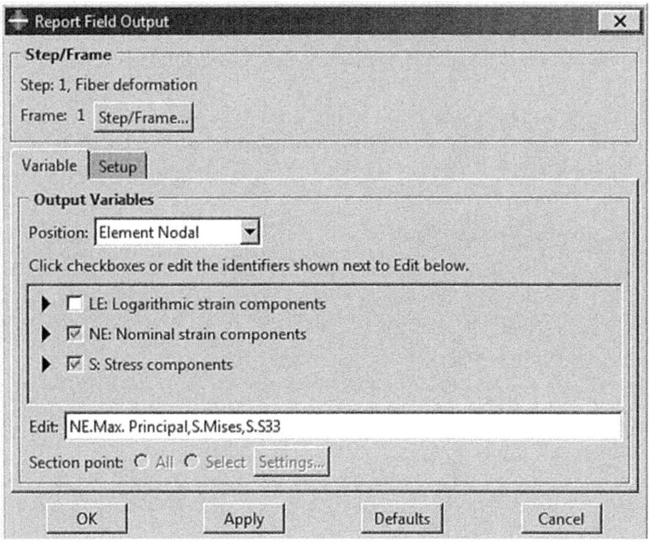

FIGURE 3.50
Report Field Output submenu—Variables. © Dassault Systèmes, a French "société européenne" (Versailles Commercial Register # B 322 306 440), or its subsidiaries in the U.S. and/or other countries.

on selected variables. In the case of current report, the variables taken into consideration are: NE Max. Principle—reflecting strain; S. Mises—reflecting Von Mises stress; and S.S33—reflecting stress but working only along *z*-axis.

The *Report Field Output* submenu allows selection of the so-called position, which is a parameter allowing to group outputs of the models according to a specific order (e.g., Integration Point, Centroid, Element Nodal, Unique Nodal, Whole Element, Element Face Nodal, Surface Face Nodal, and Element Face) (Figure 3.51).

The file titled *Unidirectional tension* having an *rtp* extension can be opened as a text document.

Only selected parameters were presented in this report as an original file of *Unidirectional tension.rpt* contains over 34,000 lines of text. The reason for it is the fact that the created model is composed of 4,331 cubic-like elements, each of them has eight nodes, and for all of them, three parameters—NE Max. Principle (strain), S. Mises (overall stress), and S.S33 (stress along *z*-axis) were calculated. Naturally, the report can be customized so that it may present the set of data that are sought. The initial part of the report contains a description, which is the name of the file, date of its creation, the type of the file, the step in the modeling process that the file refers to, etc. To limit the report's volume, only randomly selected elements and nodes together with calculated parameters were presented. The report data were sorted by Element Label. The model presented in Figure 3.52 has some Element Labels visibly presented and an element no. 1 marked. This element is also the first

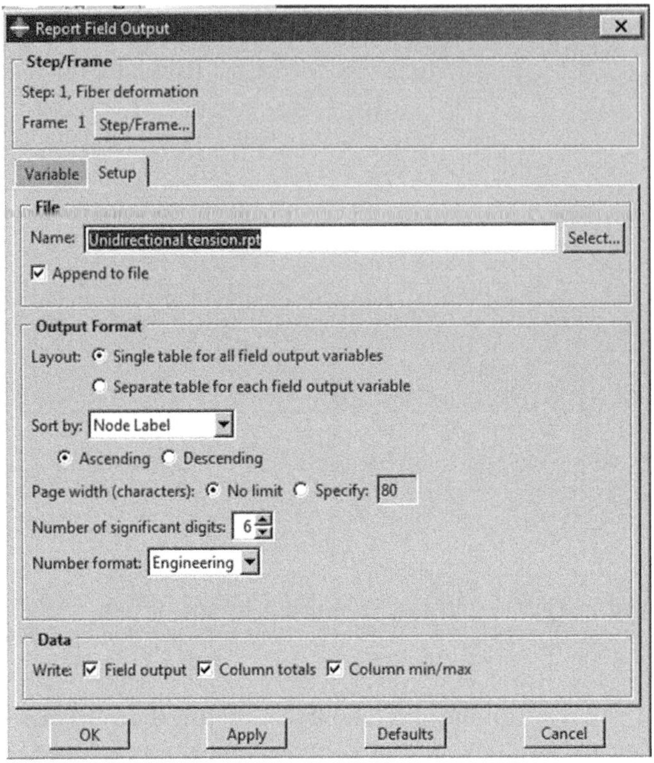

FIGURE 3.51
Report Field Output submenu—Setup. © Dassault Systèmes, a French "société européenne" (Versailles Commercial Register # B 322 306 440), or its subsidiaries in the U.S. and/or other countries.

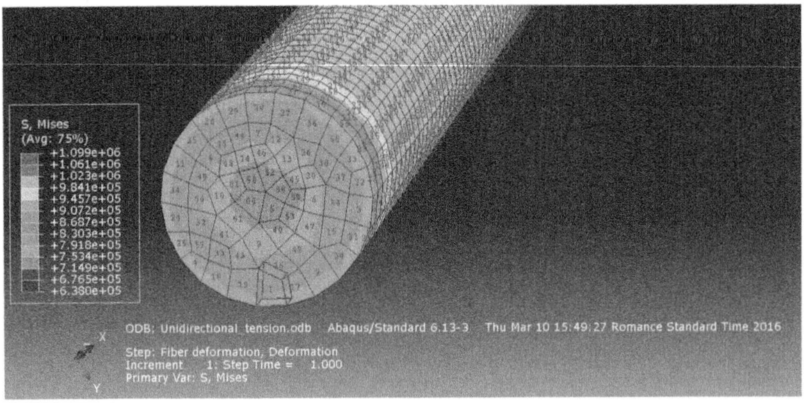

FIGURE 3.52
The model with Von Mises stress and each element of the model labeled, element no. 1 was marked. © Dassault Systèmes, a French "société européenne" (Versailles Commercial Register # B 322 306 440), or its subsidiaries in the U.S. and/or other countries.

element characterized in the report. The element has eight nodes having nos. 10, 21, 24, 40, 83, 94, 97, and 113.

Field Output Report, written Wed Mar 09 19:31:58 2016
ODB: Z:/Unidirectional_tension.odb
Step: Fiber deformation
Frame: Increment 1: Step Time = 1.000

Loc 1: Element nodal values from source 1
Output sorted by column "Element Label".
Field Output reported at element nodes for part: Nylon Fiber-1

Element Label	Node Label	NE.Max. Prin@Loc 1	S.Mises@Loc 1	S.S33@Loc 1
1	10	334.815E–06	778.007E+03	922.050E+03
1	21	340.046E–06	788.293E+03	937.516E+03
1	24	352.139E–06	804.655E+03	985.726E+03
1	40	359.236E–06	805.254E+03	1.02676E+06
1	83	347.424E–06	844.215E+03	902.494E+03
1	94	353.358E–06	857.688E+03	918.483E+03
1	97	367.929E–06	883.742E+03	932.294E+03
1	113	381.584E–06	903.037E+03	932.294E+03
2	9	337.636E–06	782.871E+03	931.263E+03
2	20	339.127E–06	786.531E+03	934.390E+03
2	25	358.167E–06	803.964E+03	1.02205E+06
2	39	357.735E–06	807.816E+03	1.01379E+06
2	82	350.909E–06	851.801E+03	912.321E+03
2	93	352.445E–06	856.147E+03	915.004E+03
2	98	380.324E–06	901.238E+03	950.721E+03
2	112	377.594E–06	899.072E+03	950.721E+03
3	7	332.319E–06	779.759E+03	904.153E+03
3	19	334.204E–06	781.404E+03	912.931E+03
3	57	355.767E–06	810.490E+03	997.596E+03
3	58	354.345E–06	810.877E+03	988.913E+03
3	80	343.071E–06	839.829E+03	881.787E+03
3	92	345.390E–06	843.469E+03	890.565E+03
3	130	372.797E–06	893.766E+03	923.862E+03
3	131	369.982E–06	889.905E+03	923.862E+03
4	12	325.109E–06	766.963E+03	880.433E+03
4	22	330.180E–06	775.070E+03	898.812E+03
4	26	345.156E–06	802.246E+03	948.236E+03
4	41	355.317E–06	815.340E+03	988.857E+03
4	85	335.034E–06	822.808E+03	858.169E+03
4	95	340.814E–06	834.426E+03	876.290E+03

(Continued)

Element Label	Node Label	NE.Max. Prin@Loc 1	S.Mises@Loc 1	S.S33@Loc 1
4	99	355.596E–06	865.037E+03	881.23E+03
4	114	369.673E–06	891.469E+03	881.23E+03
5	52	346.159E–06	741.794E+03	1.04725E+06
5	54	344.942E–06	736.486E+03	1.04945E+06
5	56	344.316E–06	736.471E+03	1.04171E+06
5	71	346.280E–06	740.621E+03	1.05091E+06
5	125	382.216E–06	870.757E+03	1.08054E+06
5	127	380.724E–06	864.153E+03	1.08212E+06
5	129	380.518E–06	865.190E+03	1.07871E+06
5	144	381.496E–06	867.234E+03	1.08197E+06
4320	5163	340.198E–06	767.657E+03	970.605E+03
4320	5164	337.277E–06	757.320E+03	970.219E+03
4320	5165	341.793E–06	772.616E+03	972.313E+03
4320	5182	339.129E–06	763.007E+03	972.217E+03
4320	5236	307.703E–06	656.749E+03	932.320E+03
4320	5237	301.377E–06	638.001E+03	923.953E+03
4320	5238	308.565E–06	659.685E+03	932.063E+03
4320	5255	302.702E–06	642.388E+03	923.770E+03
4321	5161	306.088E–06	808.746E+03	983.989E+03
4321	5165	341.793E–06	772.616E+03	972.313E+03
4321	5182	339.129E–06	763.007E+03	972.217E+03
4321	5183	340.565E–06	768.456E+03	971.930E+03
4321	5234	316.618E–06	684.706E+03	932.312E+03
4321	5238	308.565E–06	659.685E+03	932.063E+03
4321	5255	302.702E–06	642.388E+03	923.770E+03
4321	5256	306.537E–06	653.758E+03	929.467E+03
4322	5159	302.837E–06	790.582E+03	978.554E+03
4322	5163	340.198E–06	767.657E+03	970.605E+03
4322	5164	337.277E–06	757.320E+03	970.219E+03
4322	5172	341.008E–06	769.635E+03	972.398E+03
4322	5232	310.154E–06	665.296E+03	925.501E+03
4322	5236	307.703E–06	656.749E+03	932.320E+03
4322	5237	301.377E–06	638.001E+03	923.953E+03
4322	5245	305.261E–06	650.305E+03	925.653E+03
4323	5153	346.952E–06	788.173E+03	977.940E+03
4323	5166	337.876E–06	760.043E+03	969.253E+03
4323	5178	344.646E–06	781.249E+03	975.850E+03
4323	5181	338.491E–06	761.187E+03	971.193E+03
4323	5226	310.392E–06	665.838E+03	927.825E+03
4323	5239	304.329E–06	647.182E+03	927.836E+03

(Continued)

Element Label	Node Label	NE.Max. Prin@Loc 1	S.Mises@Loc 1	S.S33@Loc 1
4323	5251	308.955E–06	661.624E+03	928.833E+03
4323	5254	302.341E–06	640.967E+03	924.005E+03
4324	5137	335.828E–06	738.964E+03	994.407E+03
4324	5160	335.828E–06	738.964E+03	993.242E+03
4324	5167	381.773E–06	906.824E+03	1.01194E+06
4324	5168	383.661E–06	914.202E+03	1.01325E+06
4324	5210	326.964E–06	715.748E+03	942.387E+03
4324	5233	322.050E–06	701.810E+03	933.860E+03
4324	5240	366.066E–06	830.608E+03	1.01477E+06
4324	5241	370.939E–06	845.087E+03	1.02401E+06
4325	5116	389.810E–06	943.011E+03	1.01512E+06
4325	5127	386.752E–06	930.933E+03	1.01307E+06
4325	5170	378.198E–06	893.883E+03	1.00881E+06
4325	5171	383.974E–06	914.382E+03	1.01474E+06
4325	5189	405.783E–06	943.775E+03	1.10631E+06
4325	5200	393.453E–06	909.405E+03	1.07821E+06
4325	5243	357.158E–06	805.022E+03	997.561E+03
4325	5244	368.811E–06	838.994E+03	1.01992E+06
4330	5162	339.701E–06	765.411E+03	971.622E+03
4330	5164	337.277E–06	757.320E+03	970.219E+03
4330	5182	339.129E–06	763.007E+03	972.217E+03
4330	5183	340.565E–06	768.456E+03	971.930E+03
4330	5235	304.934E–06	648.967E+03	927.418E+03
4330	5237	301.377E–06	638.001E+03	923.953E+03
4330	5255	302.702E–06	642.388E+03	923.770E+03
4330	5256	306.537E–06	653.758E+03	929.467E+03
4331	5156	304.279E–06	810.222E+03	984.683E+03
4331	5157	304.279E–06	797.637E+03	981.093E+03
4331	5162	339.701E–06	765.411E+03	971.622E+03
4331	5183	340.565E–06	768.456E+03	971.930E+03
4331	5229	318.728E–06	690.309E+03	937.440E+03
4331	5230	313.625E–06	675.414E+03	930.989E+03
4331	5235	304.934E–06	648.967E+03	927.418E+03
4331	5256	306.537E–06	653.758E+03	929.467E+03
Minimum at		301.377E–06	638.001E+03	835.064E+03
element		4330	4330	65
node		5237	5237	99
Maximum at		439.432E–06	1.09949E+06	1.12762E+06
element		243	243	4295
node		273	273	5123
Total		13.8226	34.4491E+09	34.6856E+09

The final lines of the report presents summary of all calculations, which are minimum and maximum values at element and at node for all three parameters.

- Plotting graphs

By conducting the tensile test, one obtains the stress–strain graph. The plotted graph in Figure 3.53 presents only the elastic deformation that occurs in the initial portion of a stress–strain curve for a selected element of a model of nylon fiber.

To obtain the graph, one needs to select *Tools* in the top menu bar, when being in *Results* postprocessing mode, next select *XY Data* and then *Create.* From the submenu window called *Create XY Data,* select *ODB field output* or *ODB history output* or other options if needed.

If one selects *ODB field output* and presses *Continue,* a new submenu *XY Data from ODB field* output appears, where one may actually select the set of variables that could be utilized when plotting a graph. One selects Element nodal as a Position and two parameters for which a stress–strain curve is to be created, namely *S.S33* (a normal stress) and *LE33* (a normal strain along *z*-axis), next clicks on the *Elements/Nodes* tab and *Edit Selection.* A useful option is the one, where one selects the element from the viewport, which is *Pick from viewport.* No matter the option, labeling the elements of the model first by using the appropratie icon (Common Plot Options) situated on the right side of the viewport is very helpful. Confirm the selection by

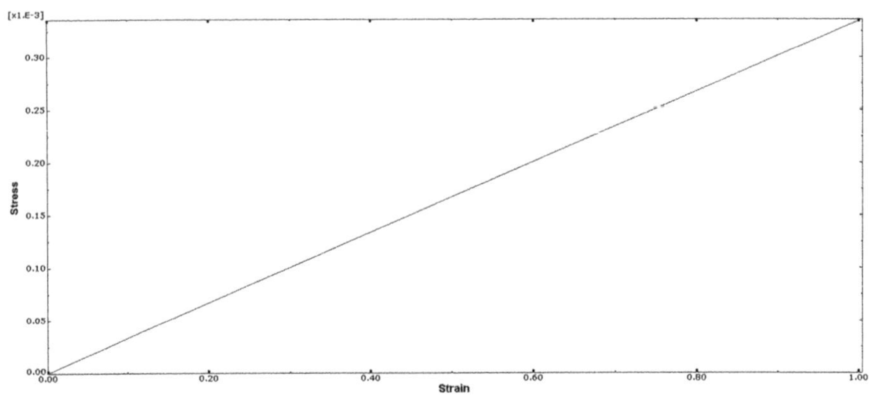

FIGURE 3.53
Stress–strain curve for a single element of the model of the fiber positioned at the edge of the model, which is not fixed, and is exposed to unidirectional deformation. © Dassault Systèmes, a French "société européenne" (Versailles Commercial Register # B 322 306 440), or its subsidiaries in the U.S. and/or other countries.

pressing *Save* button. To plot a stress–strain graph based on selected variables, go back to *Tools/XY Data/Manager/Edit.* It allows selection *Strain* on y-axis and stress on x-axis.

3.3 Summary

The tensile test analysis of a nylon fiber performed using Abaqus CAE Standard software was conducted. The obtained results were compared to the calculations performed prior to the modeling process. In terms of a displacement caused by tension applied to one of the edges of a nylon fiber, both mathematical calculations and modeling in Abaqus CAE provide the same result, which is displacement value 0.004 mm. The stress–strain curve clearly demonstrates the linear relation between these two observed parameters, which is a confirmation that the data introduced into Abaqus CAE software when preparing a model of a nylon fiber follows Hooke's law, namely stress is directly proportional to strain. As it was mentioned at the very beginning of the modeling process, only this linear part of stress–strain characterizing of the material is presented.

References

Alagirusamy, R. and Das, A. (2010). *Technical Textile Yarns - Industrial and Medical Applications.* Woodhead Publishing. Online version available at: http://app.knovel.com/hotlink/toc/id:kpTTYIMA01/technical-textile-yarns/technical-textile-yarns.

Ashby, M.F. and Jones, D.R.H. (2012). *Engineering Materials 1 - An Introduction to Properties, Applications, and Design,* 4th edn. Elsevier. Online version available at: http://app.knovel.com/hotlink/toc/id:kpEMAIPADI/engineering-materials/engineering-materials

Baboian, R. (2005). *Corrosion Tests and Standards: Application and Interpretation,* 2nd edn. (MNL 20). ASTM International. Online version available at: http://app.knovel.com/hotlink/toc/id:kpCTSAIEM9/corrosion-tests-standards/corrosion-tests-standards

Bajpai, P. (2013). *Update on Carbon Fibre.* A Smithers Group Company, Shawbury.

Belgacem, M.N. and Gandini, A. (2008). *Monomers, Polymers and Composites from Renewable Resources.* Elsevier. Online version available at: http://app.knovel.com/hotlink/toc/id:kpMPCRR002/monomers-polymers-composites/monomers-polymers-composites.

Bunsell, A.R. (2009). *Handbook of Tensile Properties of Textile and Technical Fibres.* Woodhead Publishing. Online version available at: http://app.knovel.com/hotlink/toc/id:kpHTPTTF03/handbook-tensile-properties/handbook-tensile-properties

Donnet, J.B. and Bansal, R.C. (1990). *Carbon Fibres*, 2nd edn. Marcel Dekker, New York, pp. 1–5.

Eichhorn, S.J., Hearle, J.W.S., Jaffe, M., and Kikutani, T. (2009). *Handbook of Textile Fibre Structure, Volume 2 - Natural, Regenerated, Inorganic and Specialist Fibres*. Woodhead Publishing. Online version available at: http://app.knovel.com/hotlink/toc/id:kpHTFSVNRM/handbook-textile-fibre/handbook-textile-fibre

Erhard, G. (2006). *Designing with Plastics*. Hanser Publishers. Online version available at: http://app.knovel.com/hotlink/toc/id:kpDP000002/designing-with-plastics/designing-with-plastics

Gent, A.N. (2012). *Engineering with Rubber - How to Design Rubber Components*, 3rd edn. Hanser Publishers. Online version available at: http://app.knovel.com/hotlink/toc/id:kpERHDRCE2/engineering-with-rubber/engineering-with-rubber

Getting Started with Abaqus: Interactive Edition for 6.13 version available on-line for licence holders (2016).

Hearle, J.W.S. (2001). *High Performance Fibres*. Woodhead Publishing. Online version available at: http://app.knovel.com/hotlink/toc/id:kpHPF00012/high-performance-fibres/high-performance-fibres

Honeywell Spectra® Fiber - Industrial Applications online document at https://www.honeywell-spectra.com/?document=spectra-fiber-for-industrial-applications-2&download=1 (2016).

Jose, E. and Anto, T. (2015). Analysis of tensile test of mild steel using finite element method. *Int. J. Innov. Eng. Technol. (IJIET)* 5(4): 247–251.

Kipp, Dale O. (2004; 2010). Plastic Material Data Sheets. MatWeb, LLC. Online version available at: http://app.knovel.com/hotlink/toc/id:kpPMDS0001/plastic-material-data/plastic-material-data.

Krevelen, D.W. and van Nijenhuis, K. (2009). *Properties of Polymers - Their Correlation with Chemical Structure; Their Numerical Estimation and Prediction from Additive Group Contributions*, 4th, Completely Revised edn. Elsevier. Online version available at: http://app.knovel.com/hotlink/toc/id:kpPPTCCSTB/properties-polymers-their/properties-polymers-their

Laker, R. (1987) Foam structures with a negative Poisson's ratio. *Science* 235: 1038–1040.

Mather, R.R. and Wardman, R.H. (2015). *Chemistry of Textile Fibres*, 2nd edn. Royal Society of Chemistry. Online version available at: http://app.knovel.com/hotlink/toc/id:kpCTFE0004/chemistry-textile-fibres/chemistry-textile-fibres

McKeen, L.W. (2015). *Effect of Creep and other Time Related Factors on Plastics and Elastomers*, 3rd edn. Elsevier. Online version available at: http://app.knovel.com/hotlink/toc/id:kpECTRFPE6/effect-creep-other-time/effect-creep-other-time time

Mills, N.J. (2005). *Plastics - Microstructure and Applications*, 3rd edn. Elsevier, Oxford.

Mott, P.H. and Roland C.M. (2009) Limits to Poisson's ratio in isotropic materials. *Phys. Rev. B* 80: 132104.

Plastics Design Library Staff. (1991). *Effect of Creep and Other Time Related Factors on Plastics and Elastomers*. William Andrew Publishing/Plastics Design Library. Online version available at: http://app.knovel.com/hotlink/toc/id:kpECOTRFP1/effect-creep-other-time/effect-creep-other-time

Rosato, Dominick V. and Rosato, Donald V. (2003). *Plastics Engineered Product Design*. Elsevier. Online version available at: http://app.knovel.com/hotlink/toc/id:kpPEPD0001/plastics-engineered-product/plastics-engineered-product

Ross, C.T.F. (1999). *Mechanics of Solids*. Elsevier. Online version available at: http://app.knovel.com/hotlink/toc/id:kpMS000005/mechanics-of-solids/mechanics-of-solids

Veit, D. (2012). *Simulation in Textile Technology - Theory and Applications*. Woodhead Publishing. Online version available at: http://app.knovel.com/hotlink/toc/id:kpSTTTA00Q/simulation-in-textile/simulation-in-textile

Wang, R. and Liu, Huawu (2011). *Advances in Textile Engineering*. Trans Tech Publications Ltd. Online version available at: http://app.knovel.com/hotlink/toc/id:kpATE00013/advances-in-textile-engineering/advances-in-textile-engineering.

4

Haptic Perception of Objects

4.1 Objective and Subjective Hand Evaluation—Feeling Textiles Against the Skin

Hand of Textiles (HoT) (often called handle or hand feel of textiles) has been defined in many ways. Depending on the author of the definition and the orientation of his or her studies, it is defined as

- the subjective assessment of a textile based on the sense of touch, as presented in Figure 4.1;
- the impressions that arise when fabrics are touched, squeezed, or otherwise handled;
- a person's assessment when feeling the cloth between the fingers and thumb (Beech et al., 1988; Ciesielska-Wróbel and Van Langenhove, 2012).

FIGURE 4.1
Subjective estimation of a silk fabric.

These and other definitions demonstrate certain freedom of word choice and approach to characterize a haptic perception of textiles. In order to make it more coherent and general in the same time, Ghent University, Department of Textiles, made an attempt to elaborate a temporary and precise definition of HoT, which refers to the process of handling textiles, excitement of the skin sensors, and mental evaluation of the impressions arising from the skin sensors' excitement.

The most detailed definition of a subjective HoT is that it is the act of feeling a textile's thickness, surface, and other physical features against the skin of the palm, which evokes the impressions related to physical features of the material perceived by the fingers and palm skin receptors and transferred neurologically to the cerebral cortex. These features (factors) are elusive, difficult to catch and understand and the sensations are momentary, also difficult to catch. (Ciesielska-Wróbel and Van Langenhove, 2012).

The main reasons why this area still arouses great interest in the industry and academia are

1. the fact that HoT is a crucial element influencing the purchase, both traditional and by e-commerce, of textiles by individuals;
2. the lack of a fast-performing solution: an instrument, system, or method to measure HoT;
3. the lack of uniformity as well as conformity in opinions on how to establish this complex measuring device or system;
4. the lack of a complex solution that takes into account all the haptic perception-related physical features: pressure, friction, and heat transfer from the skin surface in the direction of textiles;
5. the great divergence in the results of the measurements of HoT performed using different methods on the same pieces of textiles, and divergence in the results of the measurements of HoT performed using the same method (the same device) on the same pieces of textiles but with a different operator;
6. the lack of the possibility of fast prototyping in terms of the introduction of new textile items into the market by research and development departments of textile-related companies, whereby after producing a single prototype, one has to await the results of HoT before being able to proceed with other improvements of the prototype.

Knowing the features of the produced textiles and being able to predict them could provide producers and hand analysis experts with a tool that would allow individuals to be assured of final products with the best hand feel features, which may ultimately refer also to the judgment of ergonomic or sensorial wear comfort in the final product.

When we consider the so-called traditional purchase of textiles, usually we recall our own experience when trying on different clothing. Although the cut of the selected clothing is perfect and we look good in this piece of textile, we may sometimes have a feeling that there is something wrong with the material. It itches, and it is just not nice. So, the hand feel of this almost perfect-looking clothing makes it unattractive to buy, as nobody wishes to wear scratchy and skin-irritating textiles! In this context, the prediction of HoT is crucial for the market success of textiles, especially in terms of novel raw materials and textile structures that may be successfully introduced into the market. Another example relates to e-commerce of textiles. It is estimated that huge textile brands like Zara (Zara annual report 2013, 2014), H&M (Hennes & Mauritz Group annual report 2011, 2014), Marks & Spencer (Mark & Spencer brand multi-channel purchase information, 2016), New Look (New Look brand online shopping information, 2016), and also C&A, Next, Puma, Adidas, Tommy Hilfiger generate around 40% of their annual turnover based on e-shopping alone. In the annual report of one of the big clothing market players, selling its brand in Belgium, Germany, and UK, we read that

> ...Our on-line platform receives almost 2 million visits per week, with all of our in-store range now available to buy on-line. In addition, there is an extended range of web exclusives and boutique brands not available in our stores. We are delighted with the strong on-line progress, with sales having grown year on year by 41.4%, to £52.3 million in 2011.

In the report of H&M (Hennes & Mauritz Group annual report, 2011), we read that online shopping offers an inspiring, innovative, and interactive shopping experience. H&M's website, including the new H&M Shop Online, is today one of the world's most visited websites in the fashion retail industry.

Thus, a great effort has been made to maximize the visualization of textiles and their simulations in order to present them well via the Internet so that a decision about the purchase can be made. The existing virtual solutions do not satisfy a large group of customers who need to touch and feel the textiles before making a decision to purchase. E-customers have shared their opinions on the well-known customers' platform, Debate.org (On-line customers' debate, 2016), concerning the necessity of "feeling" items before they are purchased. In this debate, 82% of consumers are in favor of feeling the item before purchase. Thus, these opinions prove the necessity of creating a new tool or method for predicting the HoT of textiles and the ability to demonstrate it virtually:

> ...There are certain parts of the shopping experience that just do not translate into cyberspace. Touching and feeling a product is an essential part of shopping. If somebody is going to buy a tablet computer, they will want to know how it feels in their hands. If they are buying clothing, they want to know how the fabric feels against their skin

> … There are still goods out there that require an in-depth look in person. You cannot try on clothing, you cannot test drive a car, you cannot taste your favourite recipe and you can't hear what that motorcycle sounds like online. So yes, while much of the research can be done online, sometimes the final decision can be made only after touching and feeling it.

4.1.1 Subjective and Objective Techniques for HoT Measurements

The subjective HoT analysis techniques are related to analysis of the users' opinions on textiles through the application of different interview and questionnaire methodologies after an appropriate presentation of textiles. These subjective analyses are usually direct methods for making HoT measurements, which means that they categorize the fabrics immediately by describing them using adjectives such as soft–hard, limp–stiff, cold–warm, smooth–rough, and so on (Behery, 2005). The most popular ways of collecting the opinions of textile customers are questionnaires asking what the person feels when he or she touches textiles. This form of questioning may involve trained groups of panelists or nonexpert consumers, or both, observing the samples and touching them, or blind tests, whereby a panelist is required to focus purely on the haptic perception. Because of the large volume of data involved, statistical analysis and artificial neural networks are then applied to draw conclusions and to create a learning machine to learn how to perceive the textiles, thus allowing prediction of the subjective hand (Meilgaard et al., 2007; Radhakrishnaiah et al., 1993; Stearn et al., 1988).

At the other pole are the objective HoT analysis techniques, which are related to analysis of selected mechanical properties of textiles that are components of a final "value" called the HoT. Usually, objective methods do not characterize the hand directly; they provide certain mechanical parameters that are considered to represent components of the hand, such as fabric stiffness and compressibility.

4.1.2 Well-Established Objective Methods for HoT Measurement

One of the oldest and most influential studies (Peirce, 1930) established the basis for many scientific works subsequently carried out in the field of HoT. It lays the foundations for the analysis of bending length and flexural rigidity. According to Peirce and his followers, it is necessary to perform physical tests on textiles (e.g., a bending rigidity test to determine the scale of stiffness of the textiles), which may then be conjoined with the sensations that an individual feels by assigning numerical values to the measurement results. We know today that this is just the basis for HoT measurement techniques, which are far greater than what has been noted so far. The most meaningful invention related to HoT was that of Sueo Kawabata and Masako Niwa (Barker, 2002; Behery, 2005, Park et al., 2001).

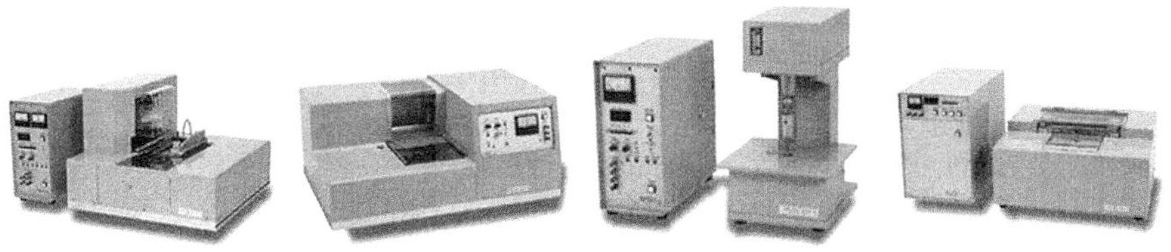

(a) (b) (c) (d)

FIGURE 4.2
The KES system—the set of four instruments and the automatic controllers necessary to estimate HoT: (a) KES-FB4-AUTO-A Automatic Surface Tester, (b) KES-FB2-AUTO-A Pure Bending Tester, (c) KES-FB3-AUTO-A Automatic Compression Tester, and (d) KES-FB1-AUTO-A Tensile and Shear Tester (http://english.keskato.co.jp/products/).

They established the Hand Evaluation and Standardization Committee (HESC) to create the so-called Kawabata Evaluation System (KES) (KES system, 2016). They published standards incorporating samples of men's winter suiting fabrics, with an overall fabric hand called the Total Hand Value. The attributes chosen by the HESC are koshi (stiffness), numeri (smoothness), and fukurami (fullness and softness). KES is manufactured by the Kato Tech. Co. of Kyoto and measures the physical, mechanical, and surface properties of fabrics using four separate instruments presented in Figure 4.2. The KES system has been successfully applied in the analysis of many kinds of textiles.

It has been proven that even a small detail in the structure of the material or its finishing can have a considerable effect on the bending properties of woven fabrics and knit fabrics, and thus on HoT (Gibson and Postle, 1978; Hallos et al., 1990; Yokura and Niwa, 2003). In order to overcome certain limitations of KES, a new set of devices called Fabric Assurance by Simple Testing (SiroFAST, 2016) system was developed by the Division of Wool and Technology at the Australian Commonwealth Scientific and Industrial Research Organization to meet the industry's need for a simple fabric performance tester. The SiroFAST system was developed initially to measure fabric performance in a garment and the appearance of the garment during wear. It turned out to provide output similar to KES from the measurements of fabric samples so that the results of the tests performed on the fabric samples may be used to estimate HoT. Although the measurements are relatively simple in comparison with KES measurements, the interpretation of the results is unfortunately equally complex. SiroFAST consists of three instruments and a test method: SiroFAST-1 is a compression meter that measures fabric thickness (corresponds to KES-FB3), SiroFAST-2 is a bending meter that measures the fabric bending length (corresponds to KES-FB2), SiroFAST-3 is an extension meter that measures fabric extensibility (corresponds to KES-FB1), and SiroFAST-4 is a test procedure for measuring dimensional properties of fabric.

4.1.3 The Innovations in Objective HoT Measurement

The most contemporary and sophisticated devices for measuring HoT, which are commercially available are

1. Tissue Softness Analyzer (TSA) by emtec Electronic GmbH, Germany.

   This is a multifunctional measuring instrument for assessing the softness, elasticity, and compressibility of fabrics as well as ball burst strength, thickness, and grammage. It measures simultaneously in a single mode somewhat different properties from KES and SiroFAST. Its advantage is that it can perform tests on both regular textiles and thick products (e.g., tissue, toilet tissue, paper handkerchiefs, facial tissue, nonwoven tissue, paper towels, paper kitchen towels, and leather). One of the main differences in the methodology of test performance is that TSA, presented in Figure 4.3, is based on acoustic measurement of softness in contrast to KES and SiroFAST.
2. Fabric Touch Tester (FTT) by SDL Atlas, United Kingdom.

   FTT is an innovative instrument capable of measuring the five parameters associated with HoT in a simple 3-minute test. These five parameters are heat flux, temperature, pressure, friction, and displacement, and thus they allow the estimation of fabric thickness, compression, bending, shearing, surface friction and roughness, and

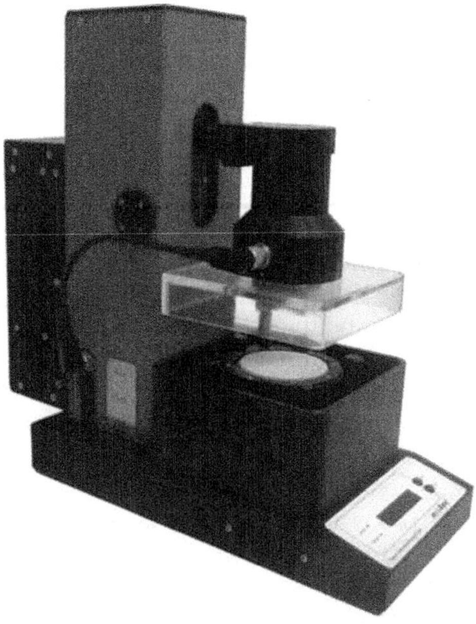

FIGURE 4.3
Tissue Softness Analyzer (https://emtec-papertest.de/en/tsa-tissue-softness-analyzer.html).

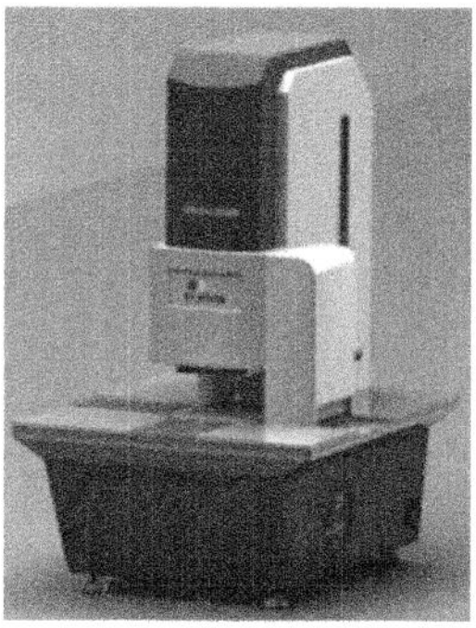

FIGURE 4.4
Fabric Touch Tester (http://sdlatlas.com/product/478/FTT-Fabric-Touch-Tester).

thermal properties. The software takes all of these data and converts them into objective measurements and graphs. A great advantage of FTT, presented in Figure 4.4, compared with old-fashioned instruments like KES and SiroFAST is the possibility of estimating the thermal aspect, which makes this device as the only complex device for measuring HoT by taking into consideration both mechanical and thermal aspects.

3. PhabrOmeter® (PO) System by Nu Cybertek, USA.

   The PO system, presented in Figure 4.5, can be used in a wide range of industries, such as textile and clothing, health care, chemical, and automobile. It has been adopted by industry textile giants from the USA, the UK, Germany, Belgium, Japan, Australia, Hong Kong, and China, with highly positive feedbacks. It offers a solution for HoT assessment and acts as a powerful tool for quality assurance and improvement of products including woven and nonwoven materials, various paper and tissues, leathers, and other sheet materials. The fact that it does not offer the estimation of the thermal aspects related to HoT is a disadvantage. Apart from softness and stiffness, as in case of KES, SiroFast, TSA, and FTT, this device can estimate a Drape Index.

4. The Handle-O-Meter (H-O-M) by Thwing-Albert Instrument Company, USA.

   H-O-M, presented in Figure 4.6, is in fact only a single module flexibility and surface fiction tester of the combined effects of flexibility and surface friction of sheeted material such as nonwovens, tissue, toweling, film, and textiles. One simply places the test sample over the slot that extends across the instrument platform and presses a test button. A penetrator beam pivots on a cam, engages the sample and forces it into the slot. Stiff materials offer greater resistance to the motion of the beam as it moves into the slot. Rough materials also exert resistance as they are dragged over the edge of the slot.

FIGURE 4.5
PhabrOmeter® System (http://phabrometer.com/).

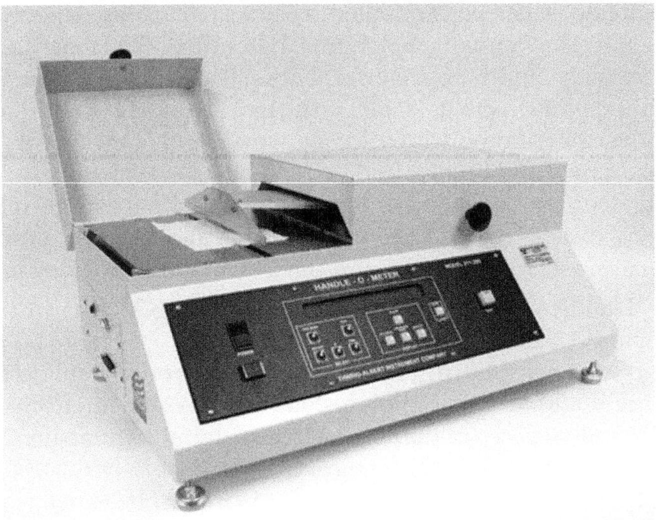

FIGURE 4.6
A Handle-O-Meter (http://thwingalbert.com/media/brochures/Brochure_HandleOMeter.pdf).

5. Polymeric Human Finger Sensor (PHFS) for friction validation.

   One of the studies performed at the School of Textile Industries, University of Leeds, United Kingdom (Ramkumar et al., 2003a,b) has led to the development of PHFS, which is an artificial human finger that can be used for measuring the frictional properties of textile fabrics. According to the study, the shape and profile of friction sensors have a profound influence on the frictional properties of textile materials. The PHFS sensor was developed from a polysiloxane compound that simulates the shape and profile of a human index finger. The polysiloxane compound was used, because its physical characteristics are similar to those of the human stranum corneum. The study found that the traction and compressibility of a human finger pad showed that the artificial finger can be used as a substitute for studying the frictional characteristics of textiles only, and thus the main limitation of this solution concerns the ability to measure only one of the attributes of HoT, which is the smoothness of fabrics. Another sophisticated detector of friction was developed by the Swiss Federal Laboratories of Material Science and Technology. The idea behind this invention was to develop detector of friction similar to the human skin. A polyurethane-coated polyamide fleece with a surface structure similar to that of skin showed the best correspondence with human skin under dry conditions (Derler et al., 2007).

   Sensory neurons innervating the skin encode the familiar sensations of temperature, touch, and pain. An explosion of progress has revealed unanticipated cellular and molecular complexity in these senses. It is now clear that perception of a single stimulus, such as heat, requires several transduction mechanisms.

   Conversely, a given protein may contribute to multiple senses, such as heat and touch. Recent studies have also led to the surprising insight that skin cells might transduce temperature and touch. To break the code underlying somatosensation, we must therefore understand how the skin's sensory functions are divided among signaling molecules and cell types (Lumpkin and Caterina, 2007).

6. Finite Element Skin Models

   Under this topic, one presents the idea/approach, which was presented by several different teams all over the world. This idea is to create a reliable and realistic skin model being a part of the body and functioning similar to the real skin (Dandekar et al., 2003; Tada et al., 2006). A deep analysis of the reaction of the skin on external stimuli is only possible by the creation of skin models. The meaningful contribution into the skin compliance description is a merit of Massachusetts Institute of Technology (MIT), Department of Mechanical Engineering, and The Research Laboratory of

> Electronics Touch Lab., which is devoted to robotic studies, developed a 3D finite-element model of human fingertips with realistic external geometries (Dandekar et al., 2003). By computing fingertip model deformations under line loads, they showed that a multilayered model matched previously obtained in vivo data on skin surface displacements. Following the experience of MIT, the mechanical and simplified thermal modeling of skin and textiles using a finite element model (FEM) has been proposed by UGent-DoT as a remedy for the current difficulties in realistic studies of HoT. The superiority of this study to MIT's studies is related to a more realistic modeling based on magnetic resonance imaging (MRI), the analysis of both skin layers and the finger. What is more, the scale of the indentations of the skin, which in the case of MIT studies was greater and easier to estimate, was smaller in fabric-related studies of UGent-DoT and, finally, basic heat transfer through the skin layers into textiles direction, which was not taken into consideration at MIT, was investigated at UGent-DoT, but only partly. A simplified solution of cylinders imitating warps and wefts in a woven textile was applied. The methodology is briefly presented in Figure 4.7.

The main aim of this study was to provide a highly repeatable, fast-performing, and fault-proof methodology for the measurement of HoT. Thus, it was decided to create 3D models of the human skin and finger by application of FEM, working with the Abaqus program, to gain a deeper understanding of the human sense of touch with reference to textiles.

A further modeling process should consider precise and accurate human skin and finger geometry as well as textile elements, their structure, and their components. The existing reference materials support creation of the

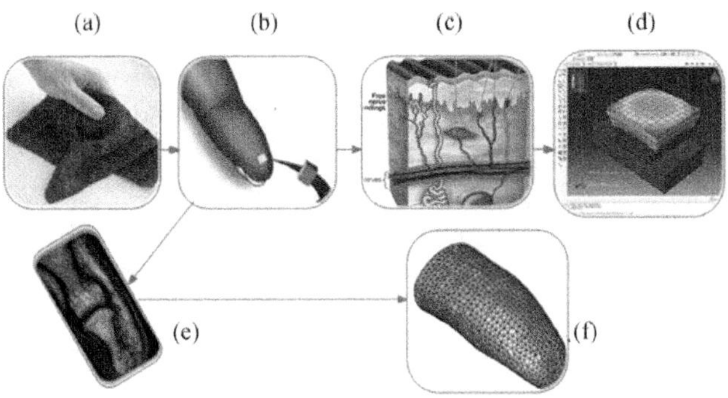

FIGURE 4.7
(a) A regular haptic perception of textiles, (b) a human finger and a skin section, (c) a skin section, (d) FEM of the skin section and textile cylinders in the background (simulation of contact of textiles with the skin), (e) MRI of the index finger, and (f) FEM of the index finger—a volumetric mesh.

full characteristic of skin receptors, so the formulas governing their activation and functionality can be elaborated and ascribed to the receptors modeled in the skin.

4.2 Model of Skin Touching an Object

The sensations that people may experience when touching an object are very rich and vary greatly depending on the person, the features of the object, the part of the skin in contact with the object, and how strongly the object is pressed against the skin. Since the skin is a complex organ, it is very difficult or even impossible to model all of its characteristics, including its precise structure and perceptive abilities. Additionally, a sensation from an object is very difficult to model from a neural point of view, due to the fact that the overall mechanism of sensation has not yet been fully characterized. In view of these obstacles, it is suggested that a simplified type of modeling is performed that takes into consideration only mechanical and/or thermal aspects, without involving the imitation of signal processing.

In this tutorial, a model is created that simulates a finger touching a piece of textile. It is sometimes difficult to generate a large textile model, due to the number of details that this textile element involves and the computational cost that is usually incurred by a detailed model. Hence, this model of the skin section and the piece of textile to be touched is reduced to two simplified elements: a skin section and a cylinder representing a piece of yarn. This model corresponds to human skin in contact with a piece of textile. The skin is pressed onto a textile element and then slides over it, reflecting a real-world situation.

As in the previous chapter, one starts by *creating a geometry*, and this is followed by deciding on which material properties should be utilized and assigning these material properties to the geometry. It is advised to follow the *model tree* and to complete the modeling process using *job* creation and visualization.

STEP 1: Creating the geometry of the skin and the object representing yarn.

The first step in creating the model is to sketch the parts that are assembled into the final model. Three parts need to be created:

- Skin: This is a 3D element with a height of 6.0 mm, a width of 7.5 mm, and a depth of 1.0 mm. In later steps, the skin will need to be divided in three sections: the external or epidermis layer, the middle or dermis layer, and the internal or hypodermis layer.
- Yarn: This is a cylinder with a radius of 0.75 mm and a depth of 1.0 mm.
- Load surface: This is a surface with a width of 7.0 mm and a depth of 4.0 mm.

The total thickness of the skin is 6 mm, and this is a measured parameter (Xu et al., 2008). The other dimensions were selected to ensure that the model is proportional and easy to work with.

4.2.1 The Skin Model

To create the skin, one performs the following steps:

- In the *Create Part* window, one names the part and chooses a *3D* modeling space. The material should be *deformable*, and a *solid* shape should be used that can be *extruded*. An approximate size of *200* is suitable.
- First, one sketches a 2D representation of the skin section using the tool Create Lines: Rectangle (four lines). One is free to select any starting point for drawing on the desktop; however, if a specific point is required, one can utilize the prompt area below the main screen in which the geometry is created and select a start and end point for the geometry.
- In this model, one uses *(3.75,3)* as starting point, and *(−3.75,−3)* as the opposite point. These coordinates can be set in the prompt area.
- In the *Edit Base Extrusion* window that appears next, one assigns a depth of *1 mm* to the skin part and confirms this choice by clicking *OK* (Figure 4.8).
- The skin section then appears.

In this model, the skin model is divided into three parts: the epidermis, the dermis, and the fatty layer—hypodermis. This means that the originally extruded element will need to be partitioned.

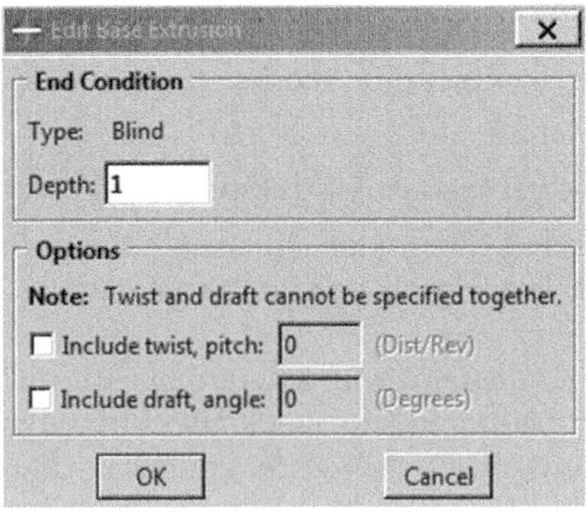

FIGURE 4.8
Editing the third dimension of the object (known as extrusion). © Dassault Systèmes, a French "société européenne" (Versailles Commercial Register # B 322 306 440), or its subsidiaries in the U.S. and/or other countries.

In the main menu bar, click *Tool—Partition* to open the *Create Partition* window. Use the instructions: *Edge—Enter parameter—Select the edge to partition—Edge parameter = 0.0167* to create a partitioning point for the epidermis/dermis separation plane. Repeat this procedure for the dermis/hypodermis layer plane, with an *Edge parameter = 0.2667* (values estimated based on those in Xu et al., 2008). The *Create Partition* window is shown in Figure 4.9.

As soon as the parameters are introduced, markers (dots) appear on the model at the location of the partition.

The partition planes are created by selecting *Tool—Partition* to open the *Create Partition* window, selecting *Cell*, clicking on *Define cutting plane*, selecting *Point & Normal* in the prompt area, selecting the partition point, and defining a normal along the edge of the modeled element. The partitioned section is shown in Figure 4.10.

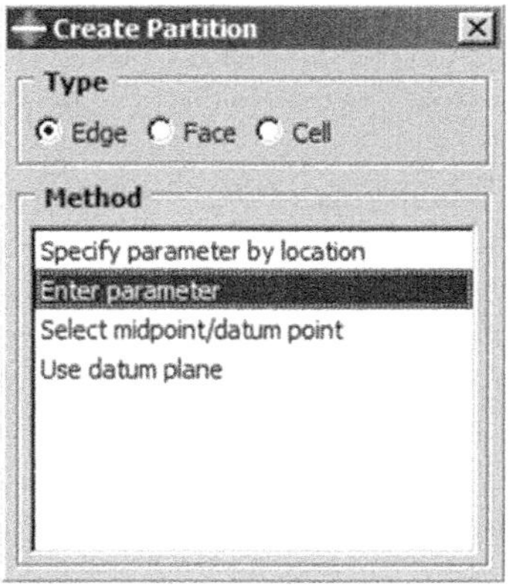

FIGURE 4.9
Creating a partition in the object. © Dassault Systèmes, a French "société européenne" (Versailles Commercial Register # B 322 306 440), or its subsidiaries in the U.S. and/or other countries.

FIGURE 4.10
A partitioned element representing skin with three layers. © Dassault Systèmes, a French "société européenne" (Versailles Commercial Register # B 322 306 440), or its subsidiaries in the U.S. and/or other countries.

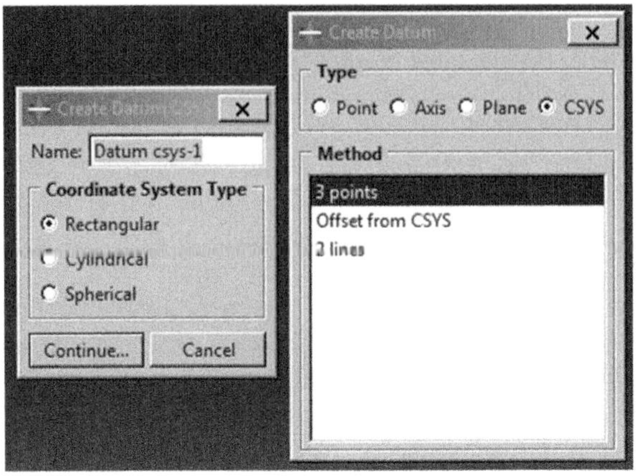

FIGURE 4.11
Creation of the CSYS datum. © Dassault Systèmes, a French "société européenne" (Versailles Commercial Register # B 322 306 440), or its subsidiaries in the U.S. and/or other countries.

Finally, a *Datum coordinate system (CSYS)* feature is added, representing a coordinate system:

- In the main menu bar, click *Tools—Datum,* and in the *Create Datum* window choose type: *CSYS>3 points,* as shown in Figure 4.11.

The *Create Datum CSYS* window appears, from which the type of the system needs to be selected. The *Rectangular* type of coordinate system was selected in this case.

4.2.2 The Yarn Model

The yarn can be created in a similar way to the skin section. The model of the yarn is a greatly simplified element and is a cylinder with dimensions proportional to those of the skin model.

- In the *Create part* window: choose *3D* modeling, *Deformable,* and a *Solid* shape created by *Extrusion.* The approximate size of the grid for drawing the model will be *200.*
- There are several different approaches to drawing a circle. The user can select an appropriate icon, *Create Circle: Center and Perimeter.* The user can type in *(0,0)* as central point and *(0.75,0)* as a perimeter point in the prompt area, and then initiate extrusion by choosing *Done* in the prompt area when requested. It is also possible to start sketching directly on the grid.
- This generates a new window, allowing us to choose the third dimension (depth) of the cylinder.

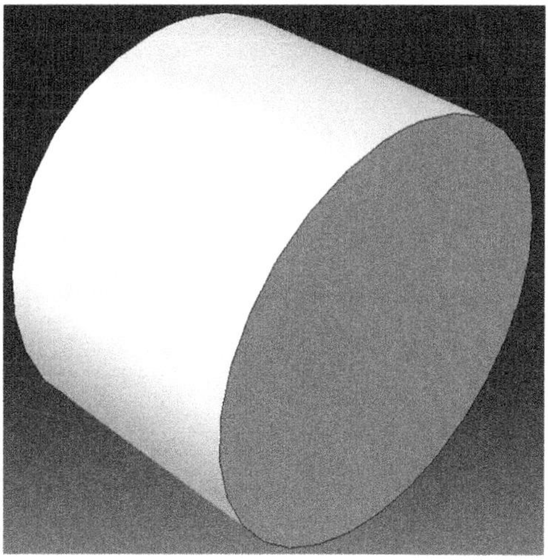

FIGURE 4.12
The constructed cylindrical object representing yarn. © Dassault Systèmes, a French "société européenne" (Versailles Commercial Register # B 322 306 440), or its subsidiaries in the U.S. and/or other countries.

- In order to keep the skin and the cylinder to the same scale and proportional to each other, one selects 1 (mm) as the depth.
- Select *Done* in the prompt area, and the cylindrical yarn will appear on the screen, as shown in Figure 4.12.

In order to be able to mesh the cylinder appropriately and ensure that it can be deformed, one can divide the cylinder into pieces using the partition process.

- In the main menu bar, click *Tools—Partition,* and in the *Create Partition* window, select: *Cell—Define cutting plane.* Next, one decides how to specify the plane and the partition. The suggested method is to select *Normal to Edge.*

The question about it appears in the prompt area after selecting *Define cutting plane.* The user may select any of these options (*Point & Normal, 3 Points,* or *Normal to Edge*), as each of these can provide the user with the desired partition; it is only the basis for the technique of this partition that differs between these three possibilities. In the current example, one selects *Normal to Edge.* As soon as this choice is confirmed, the software asks us to *Select the edge* to partition (this requires selecting the edge of the cylinder as the base for partitioning the whole cylinder). The selected edge is highlighted, and the user is asked to *Pick a point on the edge.* As soon as the point is selected, it is also highlighted, and an arrow highlighting a direction/plane of the partition is extended from the selected point. In the prompt area, *Partition definition*

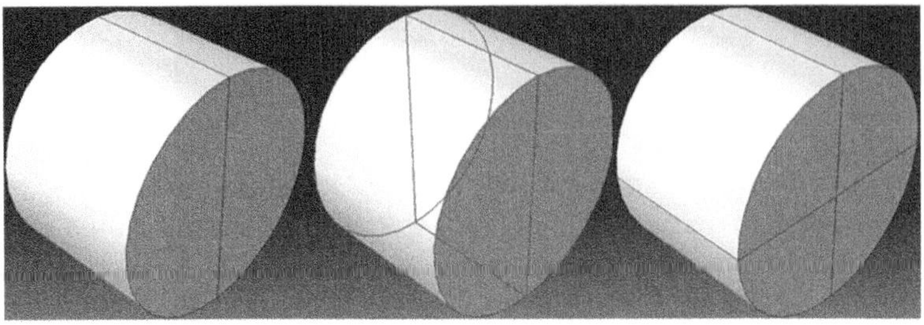

FIGURE 4.13
Stages in partitioning of the yarn model, from left to right: single partitioning; highlighting of the model edges for second partitioning; the yarn model after two partitionings. © Dassault Systèmes, a French "société européenne" (Versailles Commercial Register # B 322 306 440), or its subsidiaries in the U.S. and/or other countries.

complete is displayed with a *Create Partition* button that finalizes the creation of this partition. One may partition the cylinder many times. The presented model of the yarn was partitioned twice.

- The second partitioning plane is created in a similar way. The partitioned yarn model is shown in Figure 4.13.

Finally, a *Datum Coordination System—CSYS* should be created. This additional CSYS can be applied, for example, to define material orientations or to define an orientation for elements that form part of a larger modeling process. A thorough elaboration of the creation and utilization of CSYS is presented in the Abaqus Complete Abaqus Environment (CAE) User's Guide (Dassault Systèmes, 2014). The procedure for generating a CSYS is similar to that used to create a partition in the skin or yarn models. Go to *Tool–Datum–Create Datum*, select *CSYS* and the *3 points* method. In the new window, *Create Datum CSYS*, and choose *Rectangular*. In the prompt area, one can indicate the location at which the new Datum CSYS should be positioned. The user has the freedom to position the CSYS as required; for example, it can be positioned as shown in Figure 4.14.

4.2.3 The Load Plate

This element is created to represent the base supporting the two elements of skin and yarn. In order to create the load plate, one follows the same procedure as for creating the previous two elements. Create the part as shown in Figure 4.15. This time, one selects the *Analytical rigid* type rather than the *Deformable* type, and then selects *Extruded shell*. Using the icon *Create Lines: Connected* from the tool bar, one can draw a line starting at the central point *(0,0)* and ending at the point *(4,0)*. In the case of the current examples, one creates a part using an *extrusion depth* of 7 mm to create a load plate with the desired dimensions. Finally, a reference point is added to the load plate

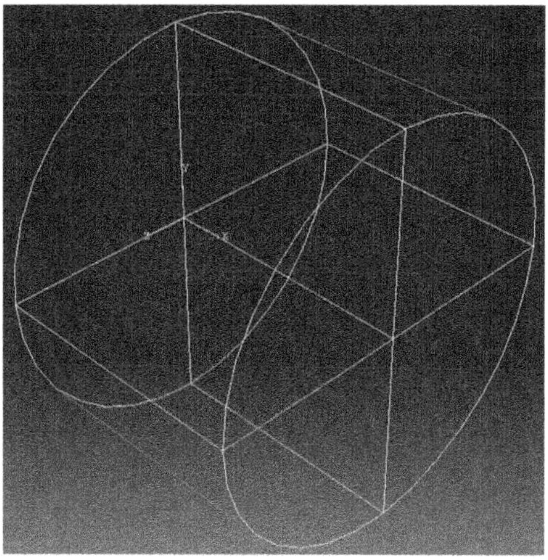

FIGURE 4.14
Positioning of the datum CSYS in the yarn model. © Dassault Systèmes, a French "société européenne" (Versailles Commercial Register # B 322 306 440), or its subsidiaries in the U.S. and/or other countries.

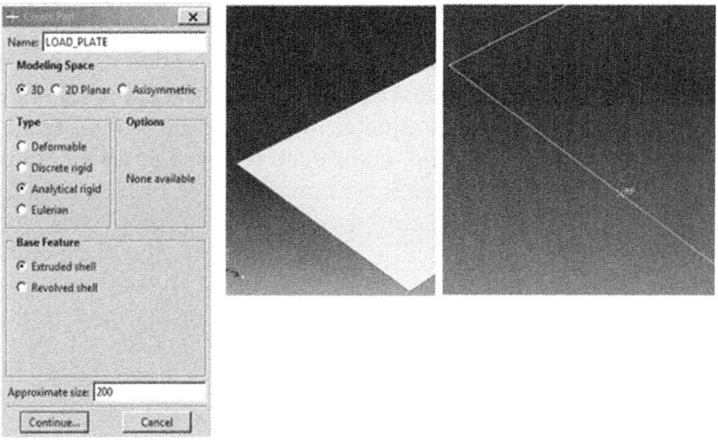

FIGURE 4.15
Creating the load plate and assigning a reference point. © Dassault Systèmes, a French "société européenne" (Versailles Commercial Register # B 322 306 440), or its subsidiaries in the U.S. and/or other countries.

by selecting *Tools—Reference point* from the main menu bar and choosing a point at the edge of the part, as illustrated in Figure 4.15.

The load plate has no specific function in this modeling, and is simply an element that holds the models of yarn and skin together and "presses" against them. The reference point located on the plate is a true reference element for the pressing and sliding motions.

STEP 2: Creating materials and assigning material properties to the geometry

In reality, human skin and the textile material are made of materials with different mechanical properties. In addition, human skin is made of different layers. The most common differentiation is that the skin is made up of the epidermis (external layer), dermis (middle layer), and hypodermis (internal layer, often described as fatty tissue). Hence, the current model needs to be composed of four different materials (plus the load plate, if one decides to retain it). In the *model tree*, one selects *Materials* and then *Create*, as shown in Figure 4.16.

The details of the mechanical characteristics of human skin can be found in the literature on this subject (Xu et al., 2008; Flynn, 2011). To edit the three layers of the skin, one follows the next steps in the *Edit material* window:

- Create a separate material property file for all four of the material characteristics involved in this modeling process: epidermis, dermis, hypodermis, and yarn. Human skin has hyperelastic properties, and this option should be selected when preparing the file.
- Select *Mechanical–Elasticity–Hyperelastic*, as shown in Figures 4.17 and 4.18.

The most important aspect at this stage is to select a suitable strain energy equation that characterizes the mechanical characteristics of the skin in the best possible way. However, various strain energy equations can be utilized to reflect the mechanical characteristics of the human skin. For some cases, the ability of the neo-Hookean equation to reflect the stress–strain characteristics of human skin is questioned or even rejected, and for some cases, it is accepted. Unfortunately, there is no hard and fast rule governing which model is most suitable for reflecting specific material properties, and this is dictated by common sense, the type of skin (or rather the place on the body where the skin is situated), and the experience of the software user, supported by reviews of the topic prior to carrying out the modeling process.

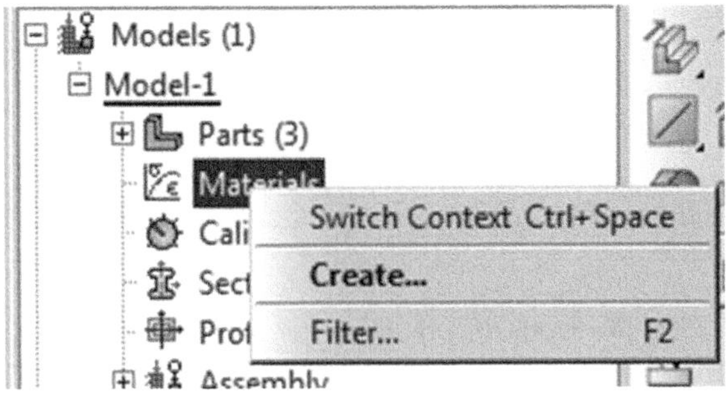

FIGURE 4.16
Creating a set of material properties for a modeled object. © Dassault Systèmes, a French "société européenne" (Versailles Commercial Register # B 322 306 440), or its subsidiaries in the U.S. and/or other countries.

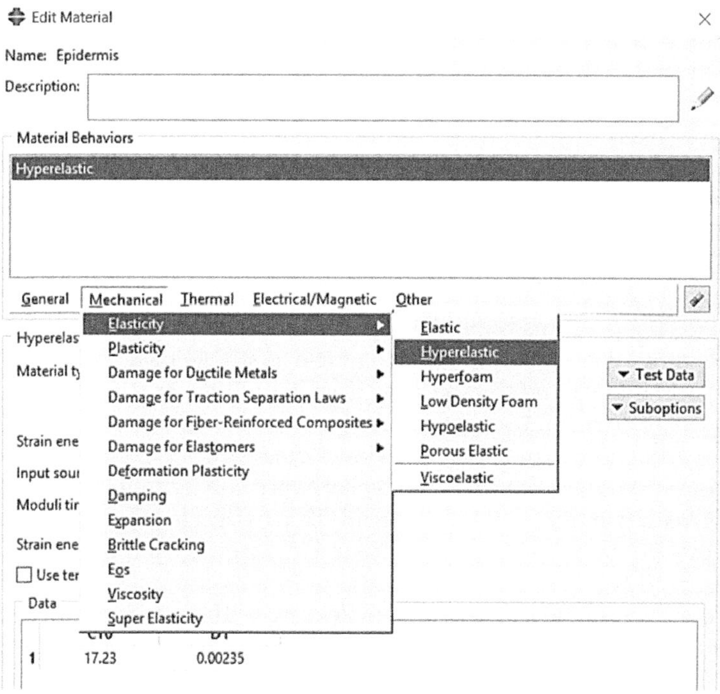

FIGURE 4.17
Editing material properties: general view. © Dassault Systèmes, a French "société européenne" (Versailles Commercial Register # B 322 306 440), or its subsidiaries in the U.S. and/or other countries.

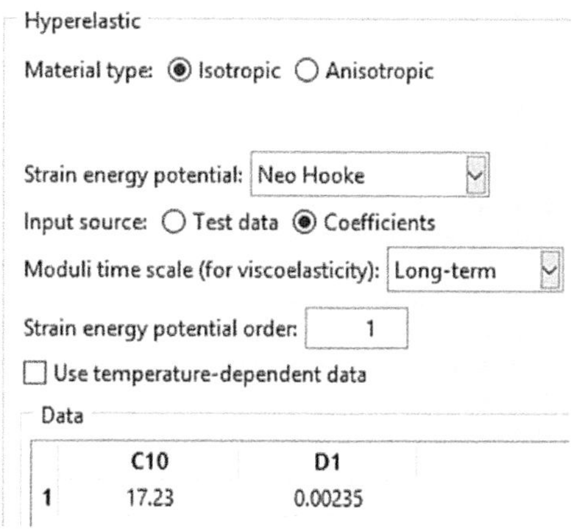

FIGURE 4.18
Editing material properties: details. © Dassault Systèmes, a French "société européenne" (Versailles Commercial Register # B 322 306 440), or its subsidiaries in the U.S. and/or other countries.

It has been accepted that, when modeling the skin, a hyperplastic material type should be used, since the mechanical characteristics of human skin are closest to this type of material (Lapeer et al., 2011). However, there are several scientific techniques that use a simplified approach in which an elastic material is used to simulate the human skin (Ciesielska-Wrobel, 2014). It is worth mentioning this it is a considerable simplification; if the strain in the material does not exceed 5%, an elastic approach may be applied, but if the strain exceeds this value, a hyperplastic approach should be used (Flynn, 2011; Groves et al., 2013). In the current model, one uses the neo-Hookean strain energy equation to reflect the mechanical characteristics of the human skin. Since users of this book are expected to be mainly textile engineering students or textile engineers, they may use available medical data concerning the skin that can be found in the literature. It is not usually possible for these users to perform tests on human skin or to verify this type of data.

Figure 4.18 shows one of the windows that appears allowing the user to select the actual data required by the software concerning the hyperelastic, neo-Hookean properties of the material. This hyperelastic property is represented by two parameters in the software, C_{10} and D_1, and these parameters are discussed further.

In order to understand what one is dealing with, one uses a definition of these materials based on the specific shear modulus.

The shear modulus of the materials (G) is defined as the ratio of shear stress to shear strain.

For materials that are isotropic (i.e. they have the same physical characteristics regardless of the direction in which they are measured), Young's modulus E can be related to G using the following formula:

$$E = 2G(1+\nu), \tag{4.1}$$

where E is Young's modulus (elastic modulus), a measure of the stiffness of a solid material in units of pressure, usually [GPa] for textile materials and [MPa] for human skin;

G is the shear modulus that reflects the reaction of the material to shear stress, in units of pressure, usually [GPa] for metals and [kPa] for human tissue; ν is Poisson's ratio, which is a measure of transverse strain to axial strain. For example, if the material is stretched rather than compressed, it usually tends to contract in the directions transverse to the direction of stretching. It has no units, and the maximum value allowed by Abaqus CAE is 0.5, while for skin tissue, Poisson's ratio is usually around 0.48.

In order to assess the C_{10} and D_1 material parameters, one needs to reach a definition and equation for the neo-Hookean energy strain, which in its simplified version is

$$C_{10} = \frac{G}{2}; \tag{4.2}$$

$$D_1 = \frac{2}{K}; \tag{4.3}$$

where K is the initial bulk modulus

$$K = \frac{E}{3(1-2v)} \tag{4.4}$$

The values of C_{10} and D_1 coefficients depend on the material.

In order to calculate these for the current examples, one uses Equation (4.1), which after transformation gives

$$G = \frac{E}{2(1+v)} = \frac{102}{2(1+0.48)} = 34.46[\text{MPa}] \tag{4.5}$$

After the introduction of values obtained from the literature for the mechanical parameters of human skin (Xu et al., 2008; Delalleau et al., 2006) into Equation (4.2), allowing us to calculate the C_{10} parameter in the neo-Hookean energy strain equation, one obtains

$$C_{10} = \frac{G}{2} = 34.46 = 17.23[\text{MPa}] \tag{4.6}$$

The K parameter is calculated similarly:

$$K = \frac{E}{3(1-2v)} = \frac{102}{3(1-2\times 0.48)} = 850[\text{MPa}] \tag{4.7}$$

allowing us to estimate the D_1 coefficient, using the formula:

$$D_1 = \frac{2}{K} = \frac{2}{850} = 0.00235\left[\frac{1}{\text{MPa}}\right] \tag{4.8}$$

which is also based on the neo-Hookean energy strain equation.

Both material coefficients C_{10} and D_1 were calculated for the single layer of human skin called the epidermis. Analogous calculations need to be carried out to assess these material coefficients for the other skin layers, the dermis and hypodermis. A summary of these coefficients calculated for all three layers of the skin is given in Table 4.1.

One may then input the appropriate values into the tables, as shown in Figure 4.18. By editing the details of the material properties, one can create a separate file for the model of the yarn. In the current examples, one uses the mechanical parameters of cotton, as this is a popular textile raw material. Thus, the suggested method is to go to *Edit Material* and to select a suitable *Material Behavior*, namely *Mechanical–Elasticity–Elastic*, as partially shown in Figure 4.17.

TABLE 4.1

Material Coeficients C_{10} and D_1 Calculated for Three Layers of Human Skin, based on human skin parameters provided in the literature

	Epidermis	Dermis	Hypodermis
C_{10}	17.23	1.723	0.1723
D_1	0.00235	0.0235	0.0235

Source: Xu et al. (2008), Delalleau et al. (2006)

- The elastic options from which one can select are shown in Figure 4.19.

The software requires us to provide values for the following parameters: *E*1, *E*2, *E*3, Nu12, Nu13, Nu23, *G*12, *G*13, and *G*23. The values were selected arbitrarily based on the information available in the literature, which varies greatly on the subject of the mechanical parameters of natural fibers. These parameters are defined as follows. *E*1 is Young's modulus measured in the *x* direction, for which one uses a value of 6,600 MPa; *E*2 is Young's modulus in the *y* direction, which is chosen as 2,200 MPa; *E*3 is Young's modulus in the *z* direction, chosen as 2,200 MPa; Nu12 $(\nu 12)$ is Poisson's ratio in the *xy* direction, chosen as 0.3; Nu13 $(\nu 13)$ is Poisson's ratio in the *xz* direction, chosen as 0.3; Nu23 $(\nu 23)$ is Poisson's ratio in the *yz* direction, chosen as 0.3; *G*12 is the shear modulus in the *xy* direction, set to 2,500 MPa; *G*13 is the shear modulus in the *xz* direction, set to 2,500 MPa; and *G*23 is the shear modulus in the *yz* direction, set to 850 MPa (Teijin, 2018; Penava et al., 2014; Srinivas et al. 2017; Gassan et al., 2001). The notations/subscripts used to characterize the parameters (e.g., 11, 22, 12) refer to the properties of the materials measured

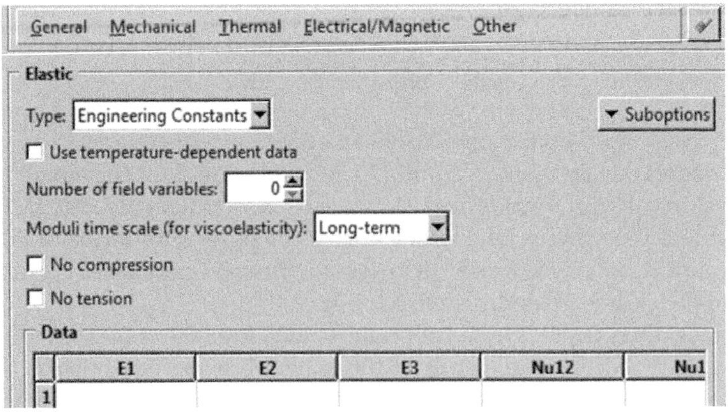

FIGURE 4.19
The parameters of the yarn. © Dassault Systèmes, a French "société européenne" (Versailles Commercial Register # B 322 306 440), or its subsidiaries in the U.S. and/or other countries.

in a specific direction of the coordinate system (CSYS). These provide information about not only the specific value but also the direction in which the property was measured.

STEP 3: Creating the section

In this step, the goal is to assign the predefined material properties to the desired sections. In order to create a section, one needs to select *Sections* in the Abaqus *model tree* (Figure 4.20a) and then select *Create*. In the *Create Section* window (Figure 4.20b), one names the section as desired; however, keep in mind that it is better to be consistent, as this makes it easier to operate with materials and sections if more than one or two elements are modeled. The name suggested here for *Section-1* is "epidermis" (other names are "dermis," "fat_layer," and "yarn_cotton"). One selects *Solid, Homogeneous* for each section in the modeling process and assigns materials to the created sections. The reason for this is that all of the materials for the skin and cotton are treated as solid and homogeneous bodies from the perspective of this modeling stage. Next, one selects *Continue* to open the *Edit Section* window, where the list of material properties created earlier is presented for selection. One connects the correct *material* to the section under construction (e.g., the material of the dermis to the section "dermis").

This step is repeated as many times as necessary. In the current model, one creates four different sections and assigns to them four different material properties. As soon as this step is complete, these sections need to be coupled to the correct regions of the different elements.

In the *model tree* in Abaqus CAE, one selects *Section Assignment* and *Create* (Figure 4.21). In the prompt area, a tip appears for proceeding further.

One *selects* the region of the skin for which one wants to assign a section, and selects *Done*. In the *Edit section assignment* window that appears, one *couples* the epidermis section to the first skin layer (the one we intend to treat as the epidermis) (Figure 4.22). If the section assignment is successful, the region changes color. This step is repeated for the dermis and fatty layer sections.

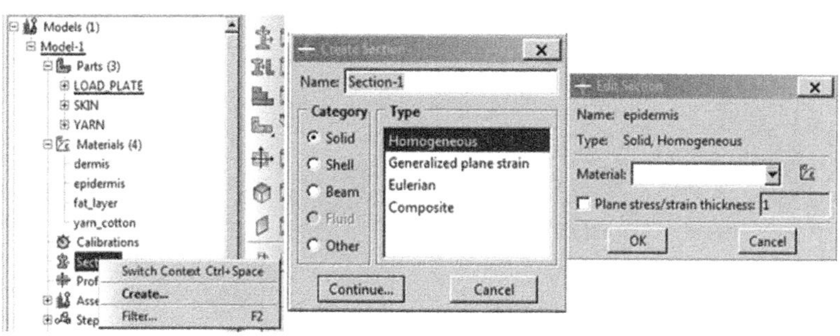

FIGURE 4.20

Defining sections. © Dassault Systèmes, a French "société européenne" (Versailles Commercial Register # B 322 306 440), or its subsidiaries in the U.S. and/or other countries.

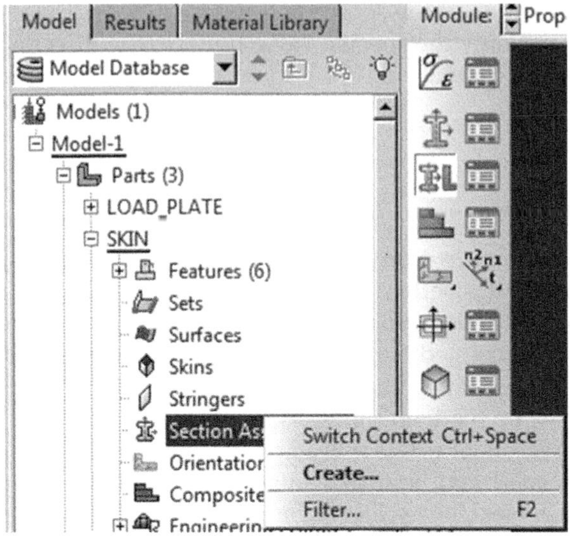

FIGURE 4.21
Assigning a section. © Dassault Systèmes, a French "société européenne" (Versailles Commercial Register # B 322 306 440), or its subsidiaries in the U.S. and/or other countries.

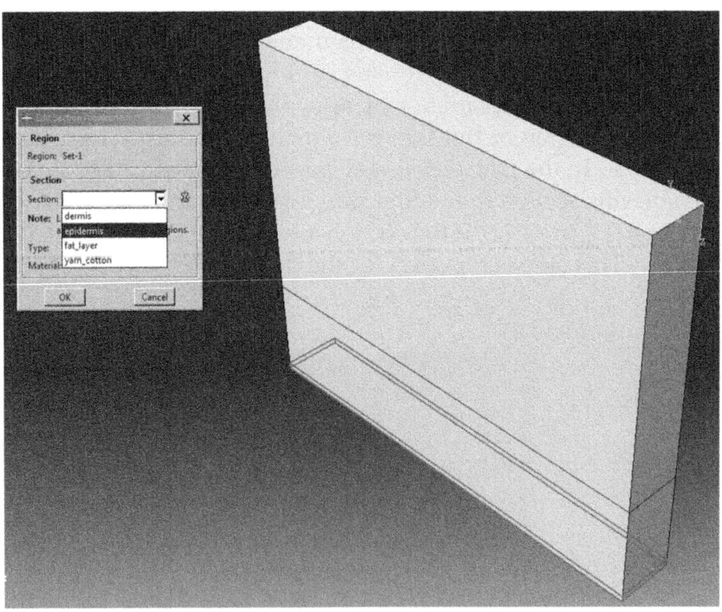

FIGURE 4.22
Assigning the epidermis section to the first layer of the skin. © Dassault Systèmes, a French "société européenne" (Versailles Commercial Register # B 322 306 440), or its subsidiaries in the U.S. and/or other countries.

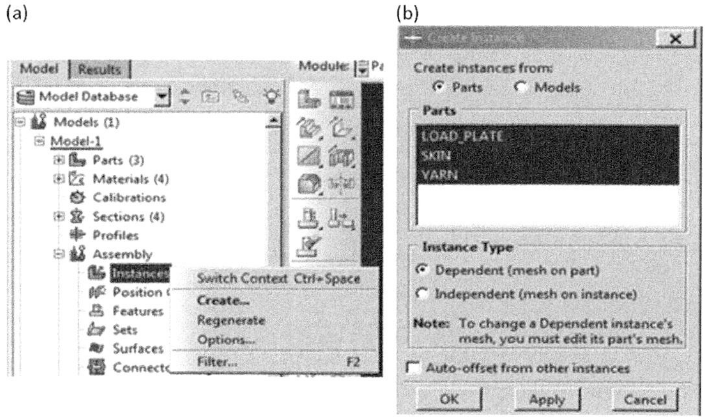

FIGURE 4.23
Combining the parts into an assembly. © Dassault Systèmes, a French "société européenne" (Versailles Commercial Register # B 322 306 440), or its subsidiaries in the U.S. and/or other countries.

For the yarn, a similar procedure can be followed; however, one needs to select the entire part and assign the yarn_cotton section to this part.

STEP 4: Defining the assembly

In this step, the parts are combined with each other to create a complete model (set of parts). In the *Model Tree*, one selects *Assembly* and then *Instances*, and then *Create* (Figure 4.23a). In the *Create Instance* window, one selects all three parts and click *OK* (Figure 4.23b). Although the part labeled SKIN was earlier divided (partitioned) into three layers, it is still a single element, and this is why only the three parts are listed.

After confirming the selection of the three parts, Abaqus CAE may position them on top of each other or in another configuration which is not desired. In this case, the program user needs to perform operations such as *translations* and *rotations* to position all the parts of the model as they are intended to interact in the completed model. The translation and rotation functions can be found under *Instance* in the main menu bar when *Assembly* is opened in the module list (these functionalities are represented by blue icons), as shown in Figure 4.24.

When the translation or rotation function is used, the prompt area asks the user to define the translation vector or the rotation vector, using a start and end point. As soon as all actions have been taken to group the parts of the model (i.e. assemble them), one obtains the assembly, as shown in Figure 4.25. Obviously, this depends on whether the user wants to place the cylinder representing the cotton yarn in the center or at the edge of the skin model. Note that there is an overlap between the yarn and the skin element, as well as a larger overlap between the yarn and the load plate element. The size of these overlaps depends on the user. In this case, these overlaps were created to reflect real-world effects and to ensure that connection points exist between the modeled parts.

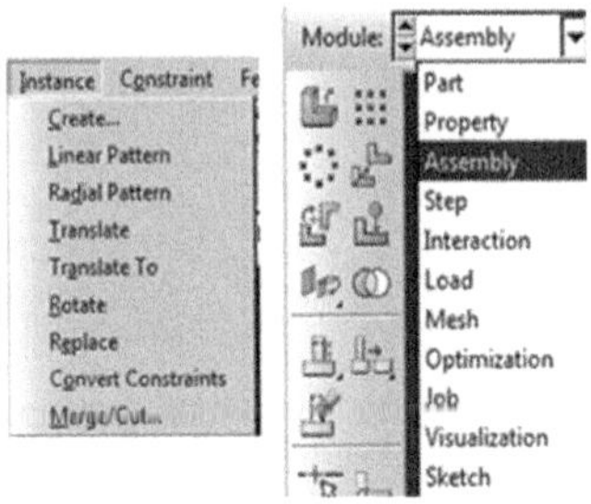

FIGURE 4.24
Tools allowing translation and rotation of instances under *Assembly*. © Dassault Systèmes, a French "société européenne" (Versailles Commercial Register # B 322 306 440), or its subsidiaries in the U.S. and/or other countries.

FIGURE 4.25
The assembly of the skin, yarn, and load plate. © Dassault Systèmes, a French "société européenne" (Versailles Commercial Register # B 322 306 440), or its subsidiaries in the U.S. and/or other countries.

STEP 5: Creation of steps

Defining steps allows us to simulate specific actions that the model can take. In our case, two steps will be created. Both of these reflect real-life situations in which contact is made with the textile by (i) touching it (pressing on it) and (ii) rubbing the skin against the textile, or in other words, moving the hand (skin) over the surface of the textile. This situation may be reversed, meaning that it is the textile that slides over the surface of the skin.

In order to define these steps, one needs to do the following:

- Select *Steps* in the *Model Tree* and select *Create*, as shown in Figure 4.26a.
- In the *Create Step* window, one names the first step "touch" and select the options shown in Figure 4.26b.
- Select *Continue* to edit the Step further. The default parameters can be maintained. Next, confirm the selection, as shown in Figure 4.27. In the *Basic* tab, make sure the *NIgeom* function is *On*, and in the *Incrementation* tab, change the *initial increment size* to 0.05 and the *maximum increment size* to 0.1, as shown in Figure 4.28.
- Use the same procedure to generate the "slide" step, but ensure that it is inserted after the "touch" step, as in the real-life situation described earlier.

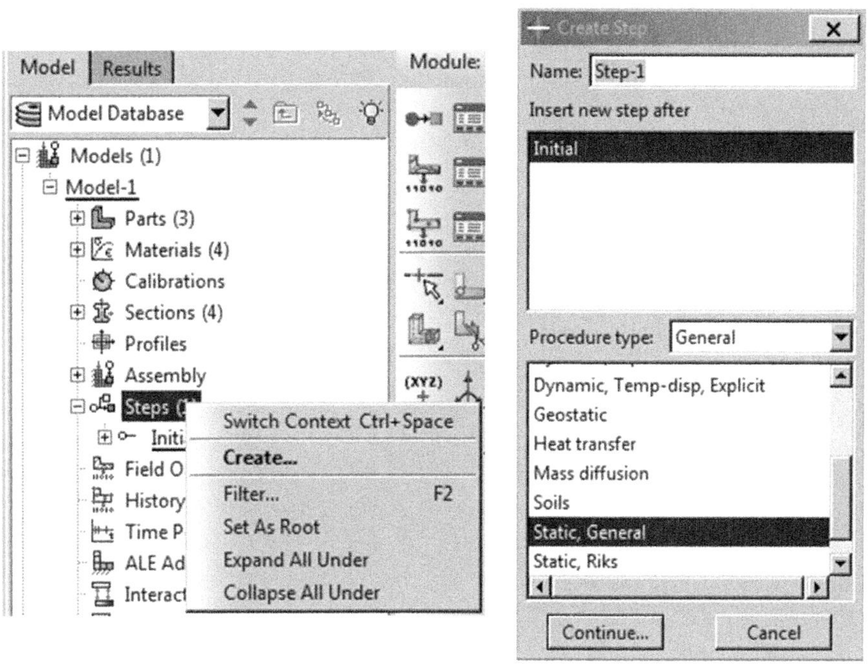

(a) (b)

FIGURE 4.26
(a) Creating a step and (b) *Create Step* window. © Dassault Systèmes, a French "société européenne" (Versailles Commercial Register # B 322 306 440), or its subsidiaries in the U.S. and/or other countries.

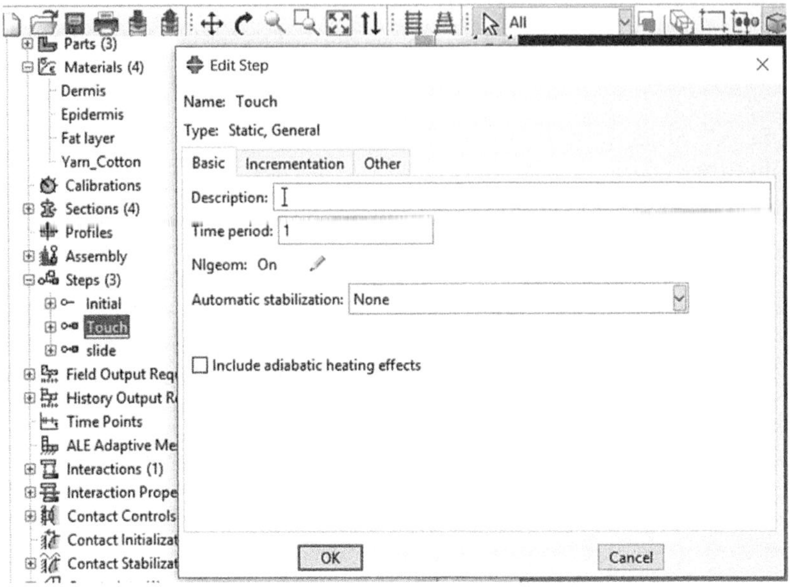

FIGURE 4.27
Editing the "touch" step. © Dassault Systèmes, a French "société européenne" (Versailles Commercial Register # B 322 306 440), or its subsidiaries in the U.S. and/or other countries.

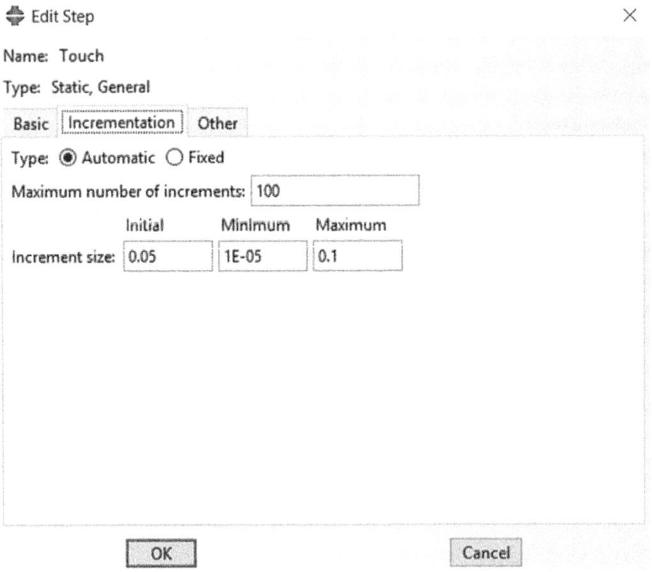

FIGURE 4.28
Editing the "touch" step: editing *Incrementation*. © Dassault Systèmes, a French "société européenne" (Versailles Commercial Register # B 322 306 440), or its subsidiaries in the U.S. and/or other countries.

STEP 6: Interaction properties

To observe any interaction between the parts, it is not sufficient to assemble them and to ensure that they are very close to each other. One also needs to characterize this interaction, mainly by defining the friction property. In order to proceed, locate the *Interaction Properties* in the *Model Tree* of Abaqus CAE and select Create (Figure 4.29). This opens the *Create Interaction Property* window.

Although the skin element was partitioned, it still exists in the model as a single element. Partitioning was done to differentiate the three layers of the skin and to assign different material properties to these, as it is expected that they will act differently under the movements defined as "touch" and "slide." Thus, the only remaining interaction to be defined is between the yarn and the external layer of the skin (epidermis) in the current model.

One takes the following steps to define the interaction between the yarn and the skin parts:

- *Name* the interaction *yarn_skin* and use the type *Contact* (Figure 4.30a) as the most suitable type of interaction between the cotton yarn and the human skin. Next, select *Continue* at the bottom of the window.

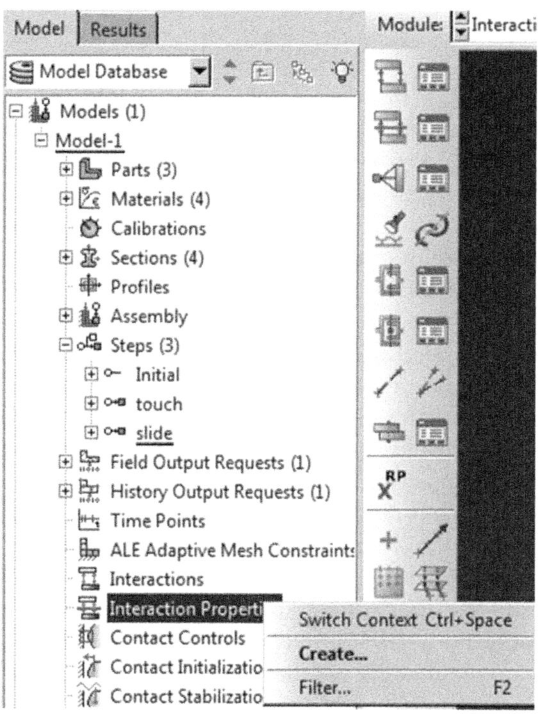

FIGURE 4.29
Defining the *Interaction Properties* of the model. © Dassault Systèmes, a French "société européenne" (Versailles Commercial Register # B 322 306 440), or its subsidiaries in the U.S. and/or other countries.

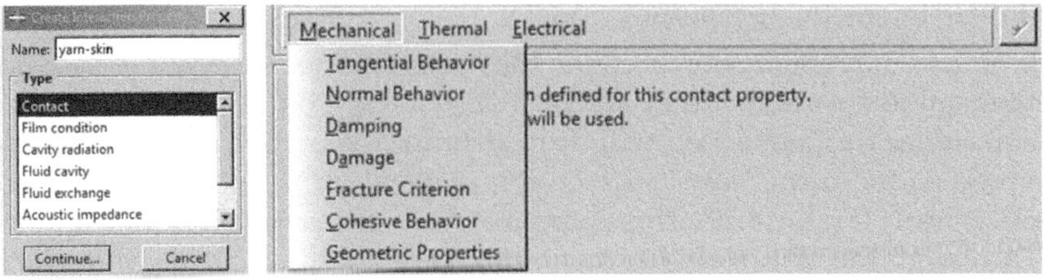

(a) (b)

FIGURE 4.30
Defining the properties of the interaction between the modeled parts: (a) contact properties and (b) mechanical–Tangential Behavior. © Dassault Systèmes, a French "société européenne" (Versailles Commercial Register # B 322 306 440), or its subsidiaries in the U.S. and/or other countries.

- A window titled *Create Interaction Property* appears that allows editing of the *Contact* property, including the tangential and normal behavior between the parts in contact. One selects *Mechanical–Tangential Behavior (TB),* as shown in Figure 4.30b. After selecting TB, one is required to select the type of contact (e.g., "frictionless" or "other").

For these two parts of the model, one selects the *Penalty* type as an interaction module. In this type, the contact force between two modeled parts is proportional to the penetration distance, meaning that some degree of penetration will occur in one of the parts. In other words, the stronger the pressure on the skin, the greater the indent in it. The system requires a *Friction Coefficient* to be entered, and here one uses a value of 0.15 (Figure 4.31a) (Vilhena and Ramalho, 2016; Ramamurthy et al., 2017; Pfarr and Zagar, 2017), which is the lowest that could be found in the literature on this subject. Since there is no agreement on the value of the friction coefficient between human skin and cotton yarn or fabric, or on the method or type of measurements, different studies have yielded different results. Additionally, the model contains an isolated cylinder representing the fiber or yarn, which is a great simplification of the modeled objects and their relation.

In addition to TB, one is required to define normal behavior, and one therefore selects *Mechanical–Normal Behavior (NB)* as shown in Figure 4.31b.

The *Constraint enforcement method* for NB should be set to *Penalty,* and the *Stiffness scale factor* should be changed to 0.01 if this is different (Figure 4.31b). The penalty method is a stiff approximation of an object-to-object contact. In reality, there is no ideal value for the stiffness scale factor. In lay terms, one can explain this factor as a stiffness quality; the contact between the parts of the model may be more rigid/hard or more elastic/spring-like, and a higher value of elasticity is closer to reality.

Confirm the selection by clicking *OK,* which will also save this interaction property.

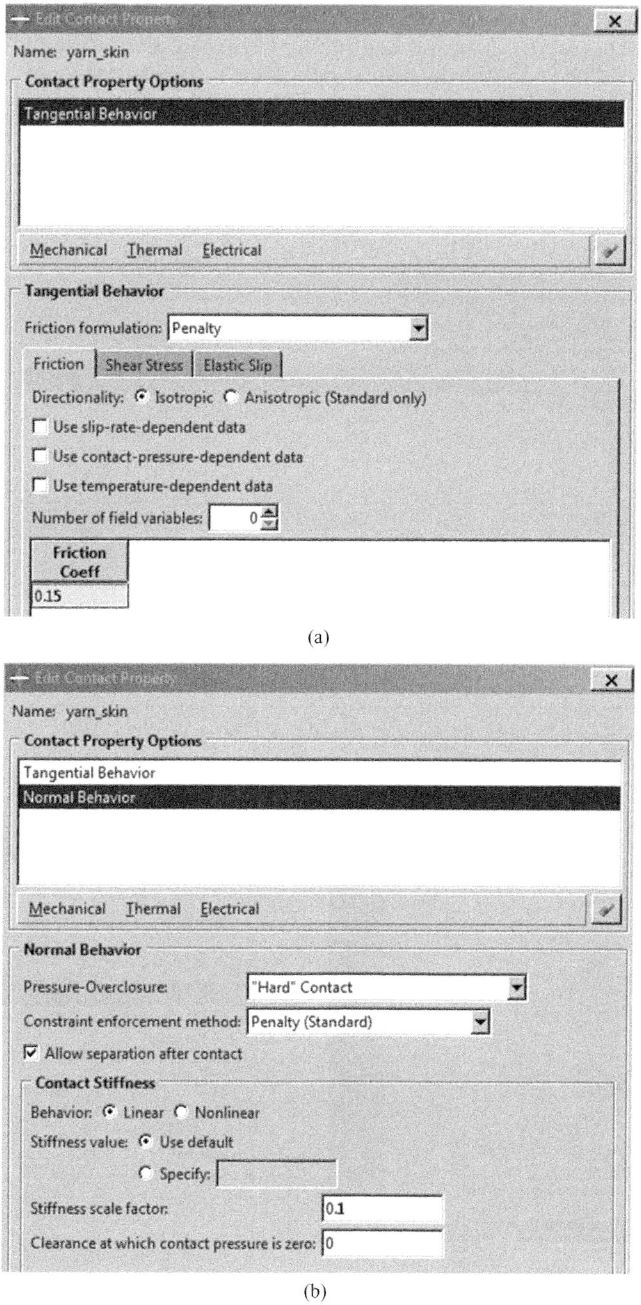

FIGURE 4.31
The parameters defining the (a) tangential and (b) normal behavior of yarn–skin interaction.

The same tangential and normal behavior needs to be set for the interaction between the yarn and the load plate. However, the name of this interaction should be changed to the name *yarn_plate* as some of the parameters (see Figure 4.32).

The interaction properties defined earlier need to be coupled to the right surfaces. To do this, one selects Interactions in the *Model tree* of Abaqus CAE and select *Create* to open the *Create Interaction* window, as shown in Figure 4.33.

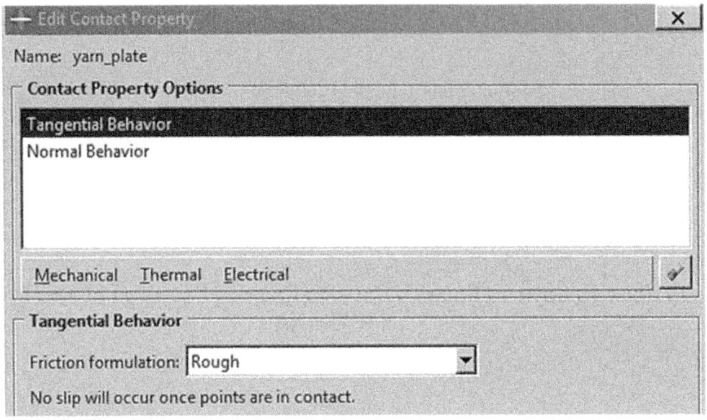

FIGURE 4.32
The parameters defining the contact property of the yarn–plate interaction. © Dassault Systèmes, a French "société européenne" (Versailles Commercial Register # B 322 306 440), or its subsidiaries in the U.S. and/or other countries.

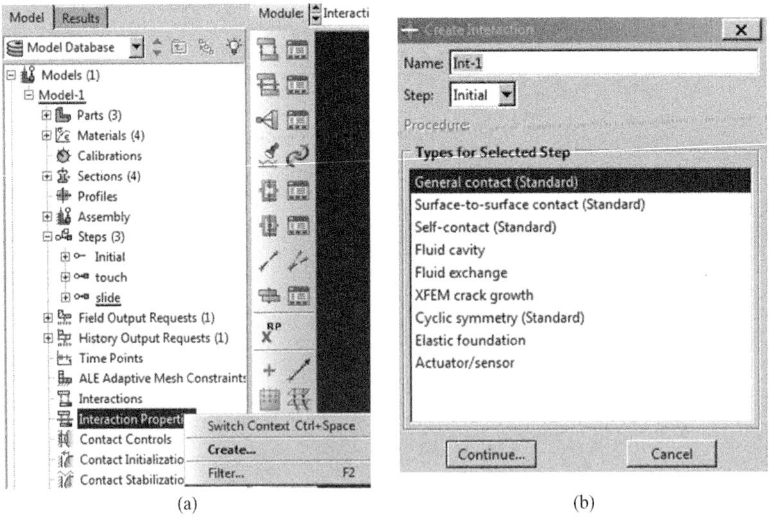

FIGURE 4.33
Creating an interaction module. © Dassault Systèmes, a French "société européenne" (Versailles Commercial Register # B 322 306 440), or its subsidiaries in the U.S. and/or other countries.

To define the yarn–skin interaction, one needs to take the following steps:

- Name the interaction and select the characteristics of the *Initial* step, using the *General Contact* type in the *Create Interaction* window. Next, select *Continue*
- The *Edit Interaction* window then appears. In this tab, the *Contact Properties* need to be defined. It is sufficient to select the property defined previously in *Global property assignment*, as shown in Figure 4.34.

One selects the *Contact Formulation* tab and goes to the pencil symbol to open the *Edit Master–Slave Assignments* window, as shown in Figure 4.35. The surfaces can be assigned by choosing the appropriate pair and confirming the choice by clicking the arrows (that also transfer the selected pair to the *Master–Slave Assignments*). For the skin–yarn interaction, one needs to ensure that the second surface chosen is the entire surface of the yarn (Figure 4.35). The yarn surface may not be in the pair table at all, as it was not defined as an independent surface. In this case, choose the symbol located below the table allowing selection of pairs to create an additional surface; after this, one will be able to continue pairing the master–slave assignments by selecting the surfaces that one wants to be in contact within the model. Finally, one confirms these selections by selecting *Done* in the prompt area. The newly created additional surface should already be present in the *Select Pairs* area of the tab, as shown in Figure 4.35.

With some differences, the yarn–load plate interaction is set up similarly. In the *Create Interaction* window, one selects the *Surface-to-surface contact* type (the second option listed in *Types* for the selected steps) and select the *Continue* key. Questions and tips are given in the prompt area. One is asked to select the *Master area*, for which one selects the *load plate*, and the *Slave area* for which one selects the *yarn surface* in contact with the load plate. This is

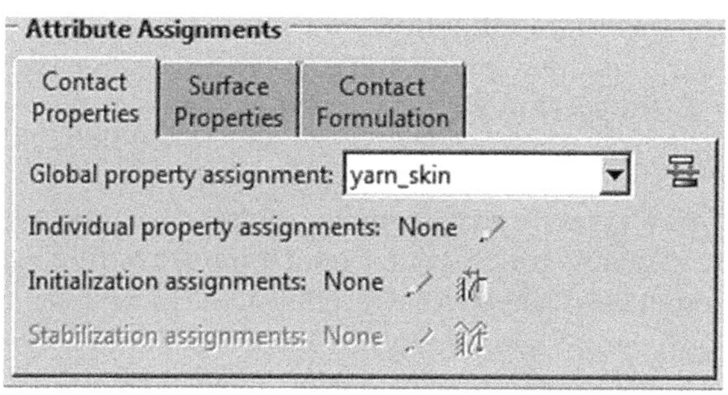

FIGURE 4.34
Assigning attributes when defining interactions between parts of the model.

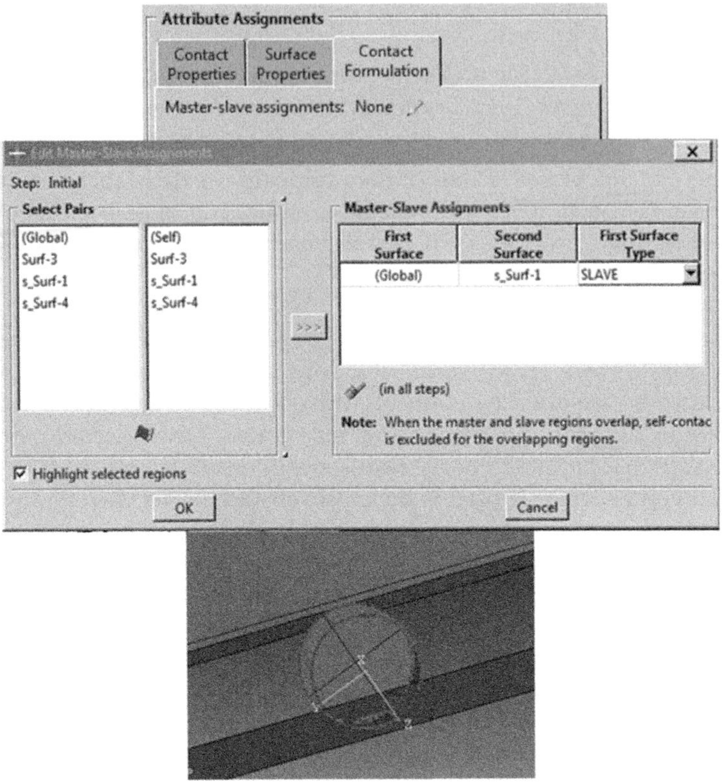

FIGURE 4.35
Editing the *Master–Slave Assignment* for the yarn–skin interaction. © Dassault Systèmes, a French "société européenne" (Versailles Commercial Register # B 322 306 440), or its subsidiaries in the U.S. and/or other countries.

confirmed by selecting the *Done* key. The *Edit Interaction* window is shown in Figure 4.36. One needs to ensure that the *Node to surface* option is set for the *Discretization method* and that this selection is linked to a suitable *Contact Interaction property,* in this case yarn_plate (Figure 4.36). These steps complete the editing of the interactions between elements of the model.

STEP 7: Meshing

Meshing involves representing a geometric object (in this model, an element) as a set of finite elements for computational analysis or modeling purposes.

All of the elements of the current model need to be meshed so that the model can be processed and analyzed and the deformations calculated. Since the parts in the model play different roles and have different shapes, they need to be meshed differently.

In the *Model tree,* one selects the element that one wishes to mesh and go deeper into the *Model tree* to select *Mesh* under the specific element that one wishes to mesh. In order to mesh the skin, one goes to the *Seed* tab in

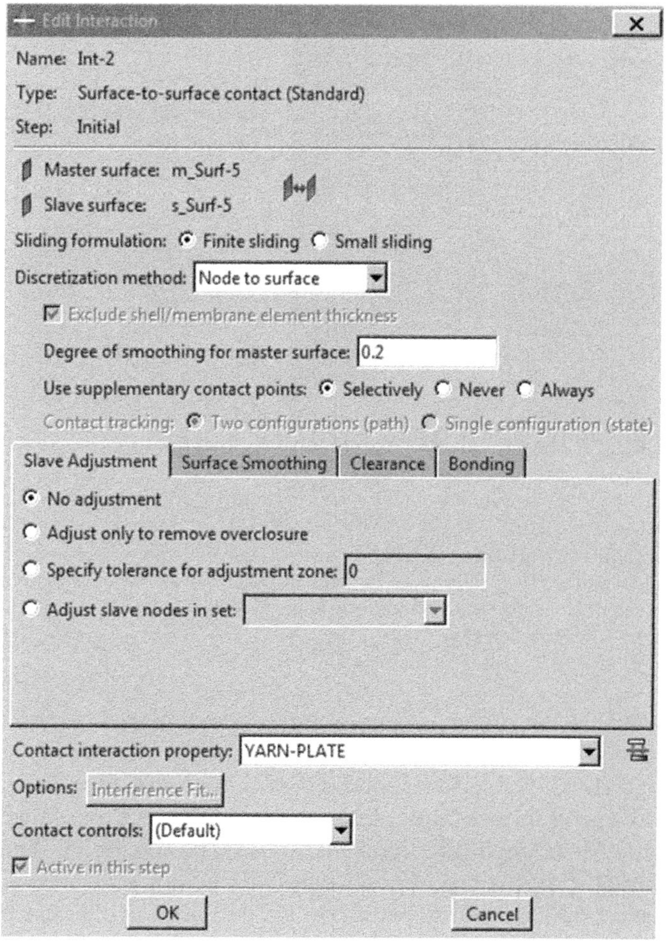

FIGURE 4.36
The *Edit Interaction* window for the yarn–load plate interaction. © Dassault Systèmes, a French "société européenne" (Versailles Commercial Register # B 322 306 440), or its subsidiaries in the U.S. and/or other countries.

the main menu bar and select *Part* (Figure 4.37a). The *Global Seeds* window appears, and one selects the *Approximate global size*, which defines the size of the single mesh element (finite element). The software calculates the optimal size of the mesh, but this can be changed depending on our requirements. In the current model, one leaves the value of 0.19 for the approximate global size for the *Skin* element. Our goal can also be accomplished by selecting *Edges* rather than *Part*. In this case, one can use one of the two methods of seeding generation and control, which is based on either size or number. The first of these leads to the same choice for the *Approximate global size* as given earlier.

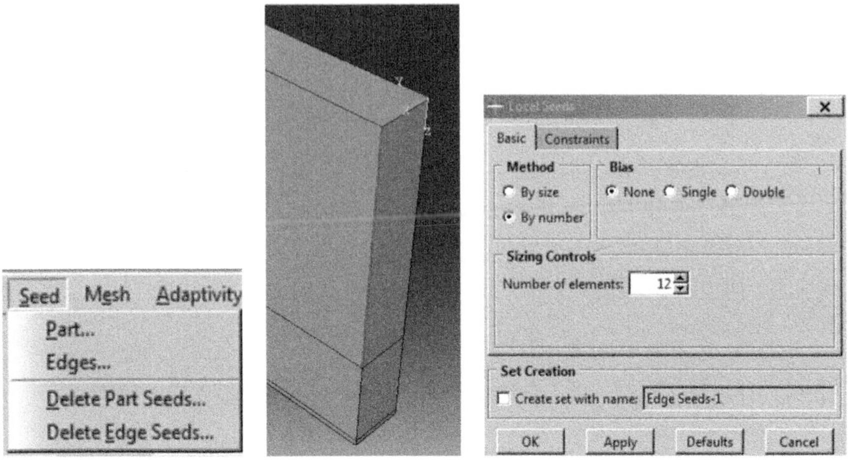

(a) (b) (c)

FIGURE 4.37
Process of meshing: (a) seeding the part, (b) selection of the edge of the model that is to be meshed, and (c) setting the number of elements in the mesh. © Dassault Systèmes, a French "société européenne" (Versailles Commercial Register # B 322 306 440), or its subsidiaries in the U.S. and/or other countries.

The second option involves generating the mesh based on the number of elements and allows us to select a desired number by which the selected edge of the part will be divided to generate a specific number of elements in the mesh. Confirming these values closes the window, and one can proceed by selecting the *Mesh* tab in the main menu bar. The *number of elements* in the current model is *12* for the hypodermis layer height (only one edge), eight for the dermis, and four for the epidermis. The selection of edges needs to be performed separately for all skin layers, and two edges per layer (length and width) should be used. As mentioned previously in this book, the mesh characteristics depend on the software user and the specific modeling needs, and it is therefore up to the software user to determine how the layers are divided.

In the next stage, one needs to select the *Element type* and the other characteristics of the mesh. Using the *Element type* window, one confirms a *Standard, Linear, 3D Stress* type of model (Figure 4.38).

There are several possibilities for selecting the element shape in this menu; however, the user is guided here to the *Mesh/Controls* menu for other options, allowing the selection of the mesh element shape.

A very important mesh and model identification is the alphanumeric code presented at the bottom of the *Element Type* window. In the current model, this code is C3D8RH, which represents an eight-node linear brick, hybrid, constant pressure, reduced integration, and hourglass control. The details of this notation are presented in the Abaqus/CAE User's Guide. For the yarn part of the model, one repeats the procedure described earlier. Figure 4.39

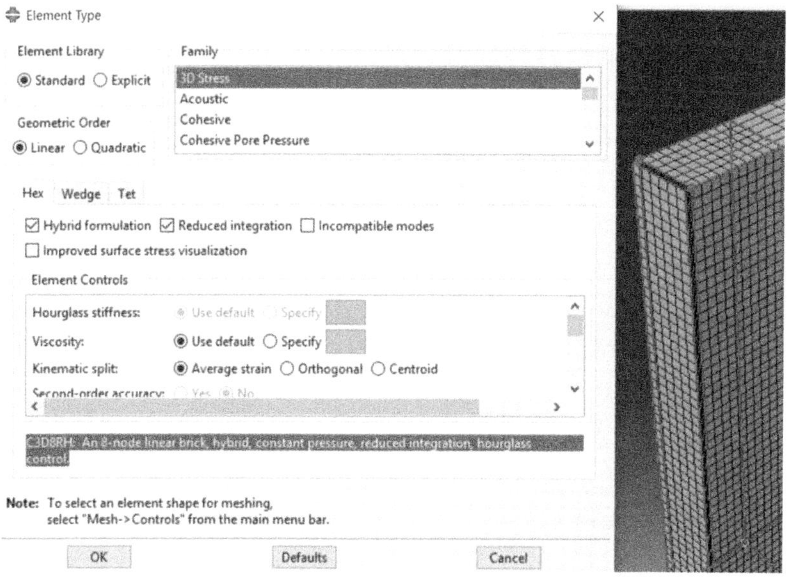

FIGURE 4.38
Selection of the mesh element type. © Dassault Systèmes, a French "société européenne" (Versailles Commercial Register # B 322 306 440), or its subsidiaries in the U.S. and/or other countries.

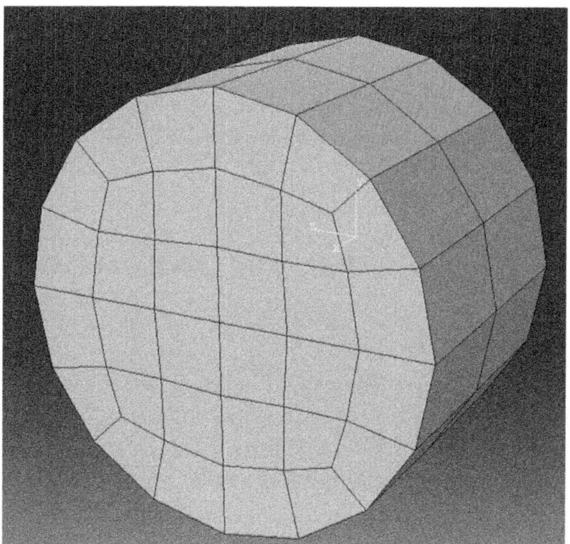

FIGURE 4.39
Meshed cylindrical part of the model. © Dassault Systèmes, a French "société européenne" (Versailles Commercial Register # B 322 306 440), or its subsidiaries in the U.S. and/or other countries.

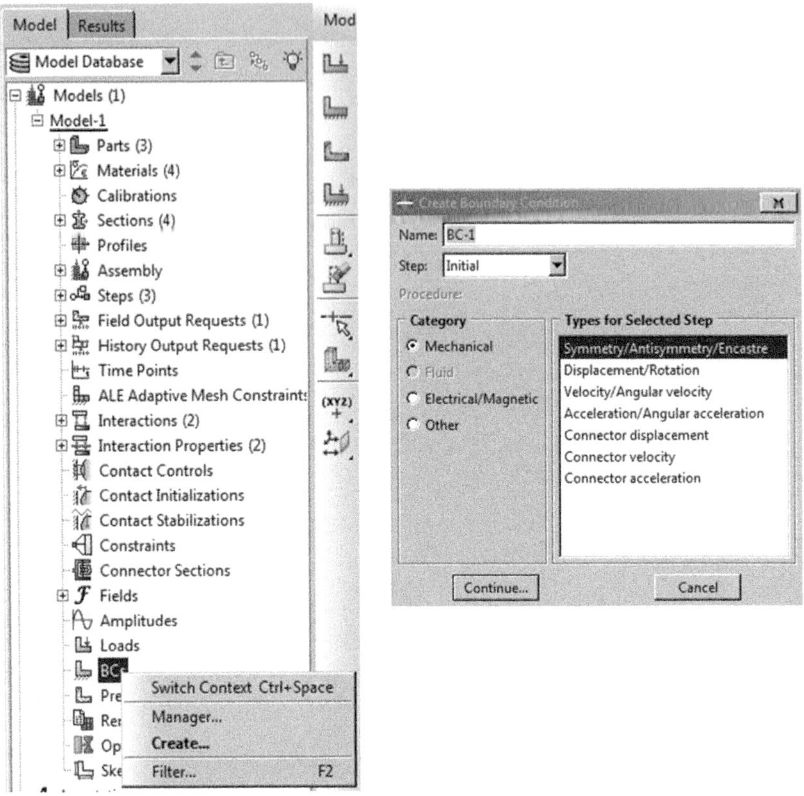

(a) (b)

FIGURE 4.40
(a) Creation of BCs for the model and (b) defining these BCs. © Dassault Systèmes, a French "société européenne" (Versailles Commercial Register # B 322 306 440), or its subsidiaries in the U.S. and/or other countries.

presents an example of the meshing of the cylindrical element representing the yarn in the model.

STEP 8: Boundary conditions

These conditions define how the model is restricted. One selects the boundary conditions (BCs) by finding them in the *Model tree* and selecting *Create* as shown in Figure 4.40a and defining them as shown in Figure 4.40b.

Four different BCs were implemented in the current model, as follows:

BC1: In the *Create Boundary Condition* window, one selects the options shown in Figure 4.41b, namely *Symmetry/Antisymmetry/Encastre* from the *Mechanical* category of procedures, to which the model is to be subjected. One confirms this selection by clicking the *Continue* key. This opens another window, *Edit Boundary Condition*, where a

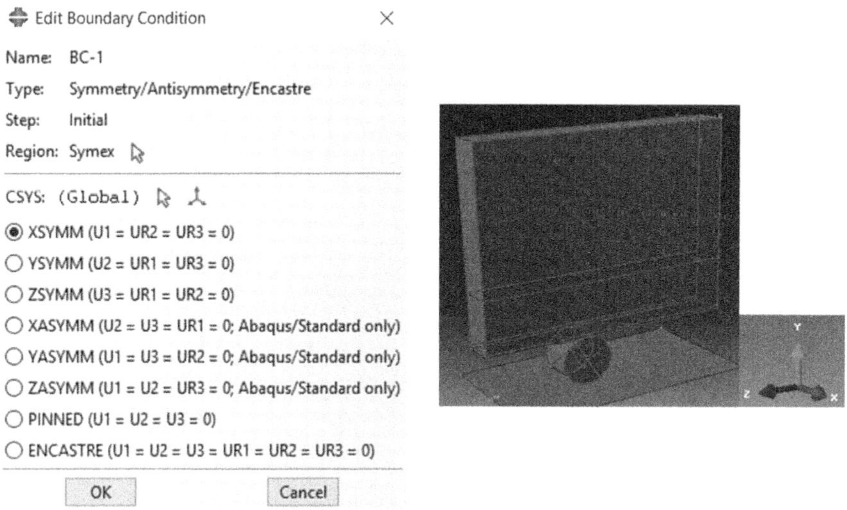

(a) (b)

FIGURE 4.41
(a) Editing BCs for the model and (b) showing BC1 on the model in red (the front and back of the skin and the yarn are marked). © Dassault Systèmes, a French "société européenne" (Versailles Commercial Register # B 322 306 440), or its subsidiaries in the U.S. and/or other countries.

summary of the previously selected options appears; this time, the user is also asked to select a CSYS datum, as shown in Figure 4.41a.

The selection in this window allows us to impose the restrictions/conditions under which the model should act. After confirming the selection of suitable BCs, the suggestion in the prompt area is to select the region(s) of the model to which these BCs will be assigned. One can select a suitable element of the model for these BCs based on how the model should act, how the model is expected to move in a specific direction, or to create a particular design. In the current model, one selects the front and back regions of the skin and yarn parts, as shown in Figure 4.41b. One expects the skin to be indented on contact with the more rigid cylinder. In the *Touch* step prepared earlier for this model, one defined the condition that the cylinder presses on the skin, meaning that an indent may appear in the skin. This effect is set to take place in one direction only, and hence BC1 limits U1, UR2, and UR3. In other words, the *XSYMM* option fixes the front and back of the skin and yarn parts in the *x* direction (the CSYS is also shown in Figure 4.41b, extreme right).

BC2: Here, one uses the same procedure but selects the side faces of the skin part and uses the *ZSYMM* option, which prevents these faces from moving in the *z* direction (see Figure 4.42).

FIGURE 4.42
BC2 marked on the skin model. © Dassault Systèmes, a French "société européenne" (Versailles Commercial Register # B 322 306 440), or its subsidiaries in the U.S. and/or other countries.

BC3: One uses the same procedure as earlier, but selects the top face of the fatty layer of the skin part and uses the *YSYMM* option, which prevents this face from moving in the *y* direction.

BC4: One selects *Displacement/Rotation* in the *Create Boundary Condition* window and then selects *Continue*. This displays a *Reference point* on the load plate that needs to be selected. One confirms the selection by choosing *Done* in the prompt area.

As a consequence, a new *Edit Boundary Condition* window appears, and this time one selects U1, U2, U3, UR1, UR2, and UR3 to add limitations to the model. Some adjustments are necessary in this BC in order to introduce the movement of the yarn with respect to the skin:

- In the Abaqus *model tree*, one navigates to *BC4* and edits the touch step. The *U3* value is changed to *−0.5* in the *Edit Boundary Condition* window, as shown in Figure 4.43.
- The slide step of *BC4* is also edited, where *U2* and *U3* are *−1* and *−0.5*, respectively.

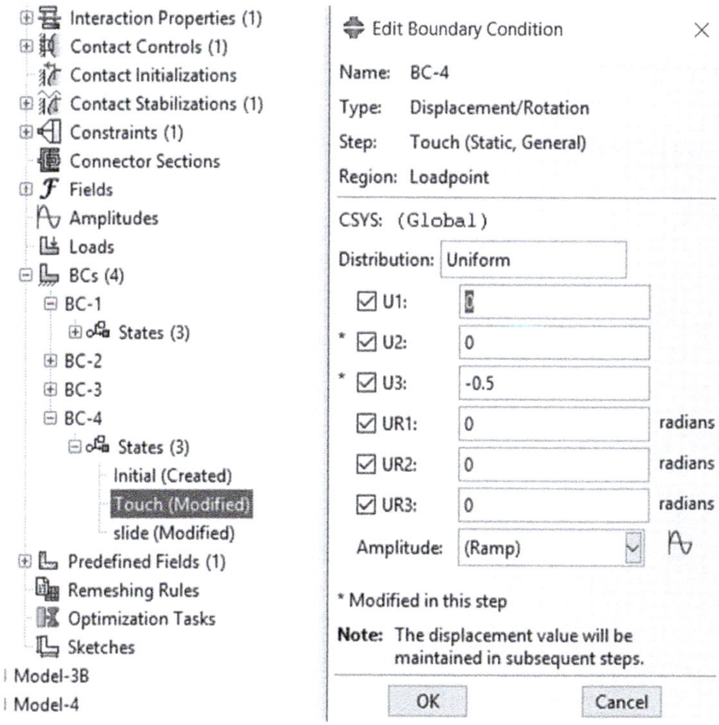

FIGURE 4.43
Editing BC4. © Dassault Systèmes, a French "société européenne" (Versailles Commercial Register # B 322 306 440), or its subsidiaries in the U.S. and/or other countries.

These adjustments represent the leeway allowed in the model when the actual movement takes place in the touch and slide steps. In other words, it indicates how far the model can move in the U2 and U3 directions when the relevant steps are initiated.

STEP 9: Creating contact controls and stabilization

Contact stabilization introduces damping to oppose the incremental relative motion between surfaces in contact (the skin and yarn in the current model). The use of these functions in the model may negatively impact the accuracy of the results but may support convergence. More information on this topic is available in Dassault Systèmes (2014) and newer versions of this document.

An automatic *Warning* is also displayed in the *Edit Contact Controls*, as illustrated in Figure 4.44c.

As explained in the Abaqus software manual, automatic stabilization is used to implement the default damping coefficient calculated automatically by Abaqus/Standard. If necessary, one may enter a value of the *Factor* by which the default damping coefficient will be multiplied.

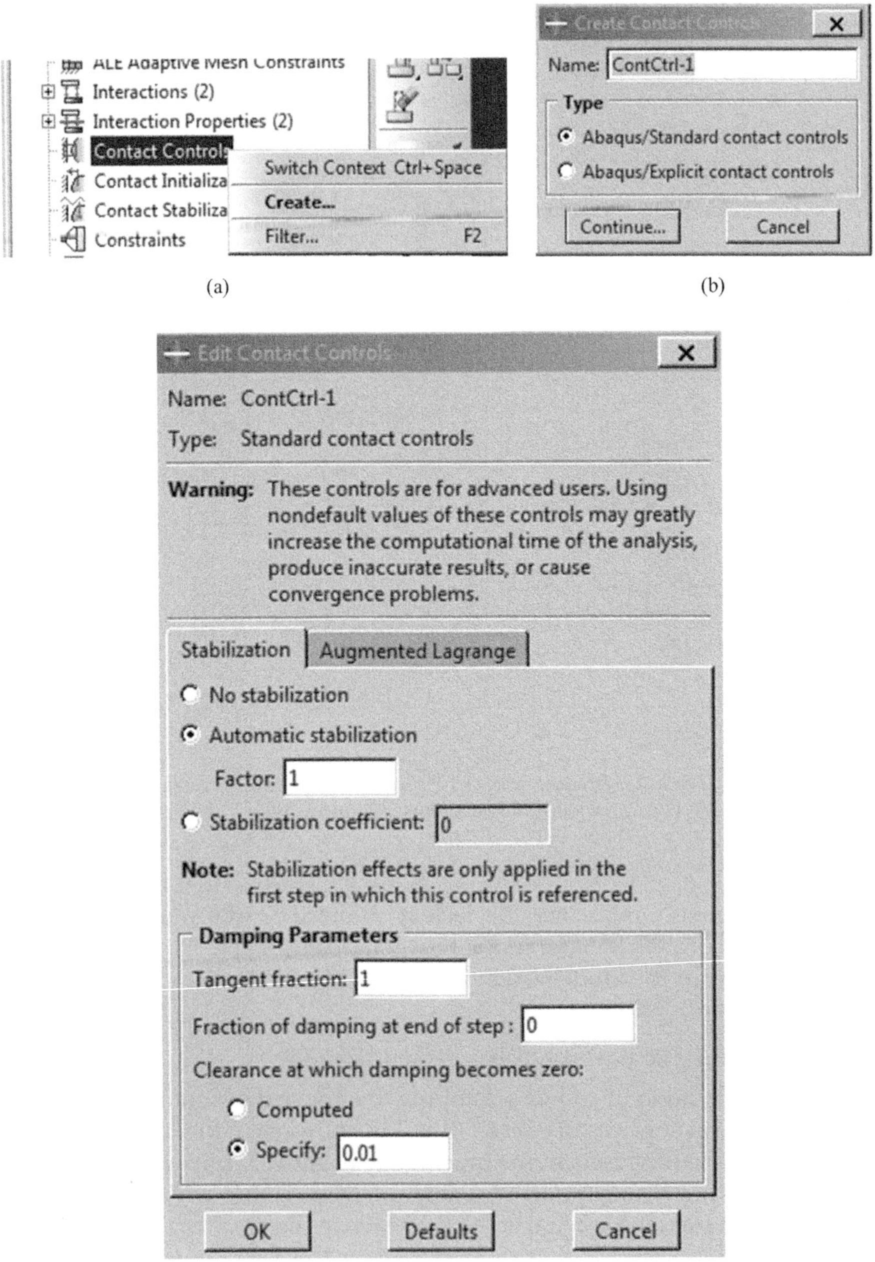

FIGURE 4.44
Introducing contact control parameters. © Dassault Systèmes, a French "société européenne" (Versailles Commercial Register # B 322 306 440), or its subsidiaries in the U.S. and/or other countries.

In the *Tangent fraction* field, one enters a value for the fraction of normal stabilization. By default, the tangential and normal stabilization are the same.

In order to add contact controls and stabilization, one needs to navigate to *Contact controls* in the *Model Tree* and select the *Abaqus/Standard Contact Control* from the *Create Contact Controls* window if this was not automatically preselected. One uses the parameters shown in Figure 4.44.

In order to introduce the contact stabilization, one selects *Contact stabilizations* in the *Model tree* and then selects *Create* (Figure 4.45a). The *Edit Contact Stabilization* window appears, allowing us to change the *Tangential factor* to *1* (Figure 4.45b).

STEP 10: Orientations

This option is used to define a local coordinate system specifically for the definition of material properties (e.g., anisotropic materials or jointed materials); for the definition of local material directions, such as the in-plane fill and warp yarn directions of a fabric material (Dassault Systèmes, 2014); for material calculations at integration points; and for the definitions of element properties. In other words, one introduces a new orientation system for a specific part(s) of the model, to inform the system/software that these parts (and particularly their materials) need to be considered in a different way from the rest of the parts in the model.

For both parts of the model (skin and yarn), one selects *Orientations* in the Abaqus *Model tree* (Figure 4.46a). As usual, the prompt area displays a comment/tip, and this time, it is a command to select a region to which a new orientation needs to be added. One selects the entire skin and yarn parts and

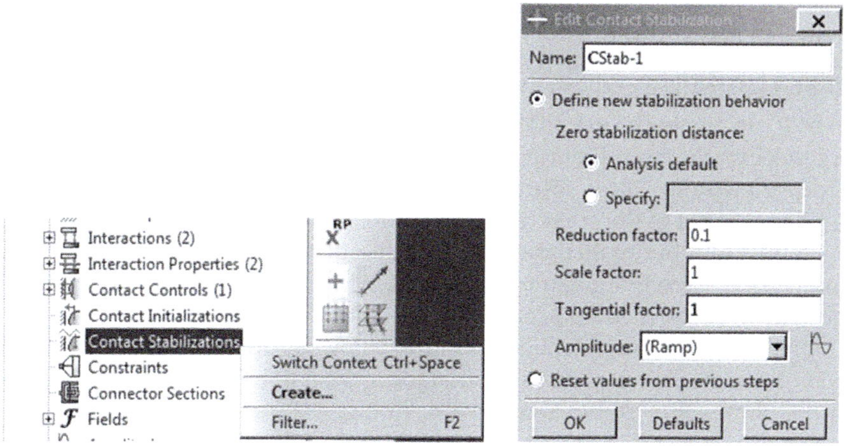

(a) (b)

FIGURE 4.45
(a) Creating contact stabilizations and (b) editing contact stabilization. © Dassault Systèmes, a French "société européenne" (Versailles Commercial Register # B 322 306 440), or its subsidiaries in the U.S. and/or other countries.

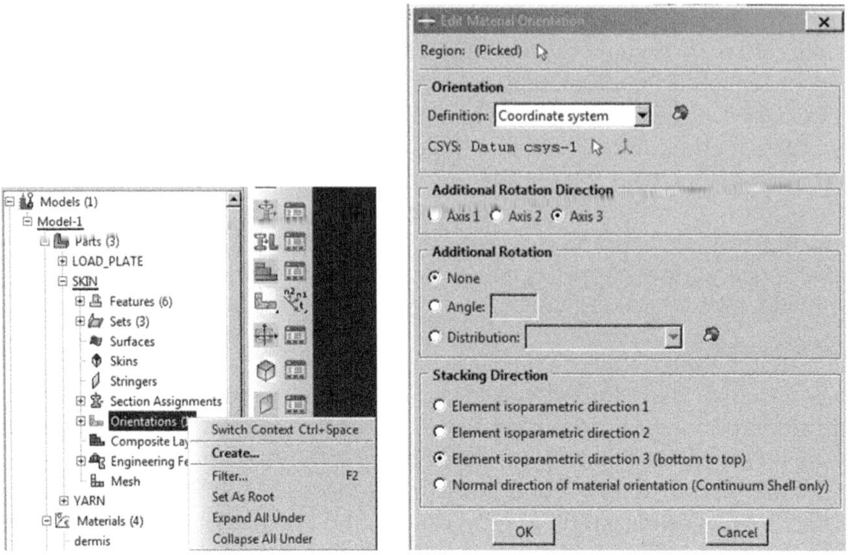

(a) (b)

FIGURE 4.46
Creating a material orientation.

confirms the selection by clicking *Done*. On the left-hand side of the prompt area, the *datum CSYS* list appears. One selects the *datum CSYS point* that was previously assigned to the skin or yarn part. The details of this selection are shown in Figure 4.46b.

STEP 11: Creating and executing the job

In order to create the job, that is, a summary of the whole modeling process, one selects *Jobs* from the *Model tree*, and gives a name to the new *Job* in the *Create Job* window. As a source, one selects *Model*, as the file to be processed is a direct effect of the creation of the model. In the same window, one selects the model number or its name. If the author of the model did not name it, an automatically assigned name (e.g., Model-1, Model-2) will be applied, and the job will be performed on Model-1 or Model-2, respectively. Next, one selects *Continue*, and the newly opened window allows us to give a specific description of the modeling process or an additional name for the model. The rest of the parameters do not require any changes. As soon as a new *Job* is created and is present in the *Model tree* under the *Jobs* tab, one can click on the right mouse button when the new *Job* is highlighted and select *Data Check*, as shown in Figure 4.47.

If the *Data Check* step is successful (comments on this or errors detected can be found in the prompt area), one can continue and select *Submit*. One has the option to improve the model before it is submitted and fully analyzed.

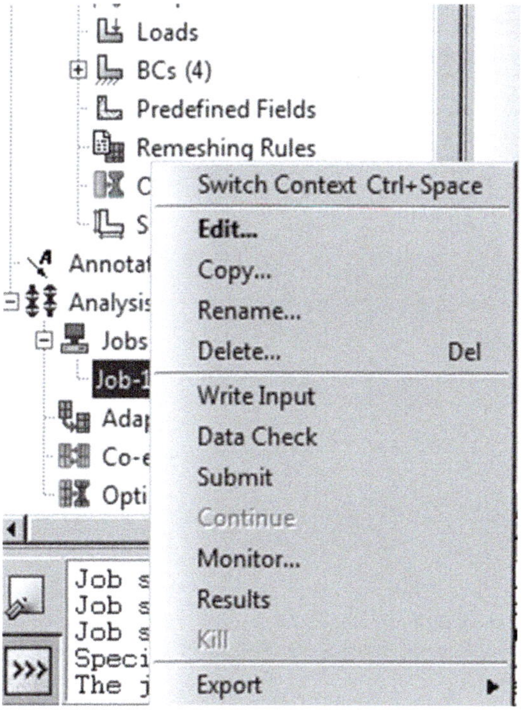

FIGURE 4.47
Running the job. © Dassault Systèmes, a French "société européenne" (Versailles Commercial Register # B 322 306 440), or its subsidiaries in the U.S. and/or other countries.

Convergence issues may arise in the model, and if so, this will be indicated after submission. The specific incrementation and processing of the model can be observed after selecting *Monitor* as shown in Figure 4.48.

STEP 12: Visualization

When the *Job* processing is successfully completed, as announced in the prompt area, one can return to the *Model tree* to select *Results*. When the software switches to the *Results module*, a completely different set of icons and options become available to the user. From the tool bar, one selects the icon *Plot Deformed Shape* to observe how the whole model will look in the case extreme deformation. Next, one can select one of the icons: *Plot Contours on Deformed Shape*, *Plot Contours on Undeformed Shape*, and *Plot Contours on both Shapes*. The plot contours in these cases depend on the parameters that are chosen for analysis when working on the model via the *Field Output Requests* and/or the *History Output Requests*, which are located in the *Model tree*. If no specific parameter has been selected, the software may use certain default parameters (e.g., S (Von Mises stress) and U (displacement)). It is worth using *Animation Options* and other related icons to observe the full animation

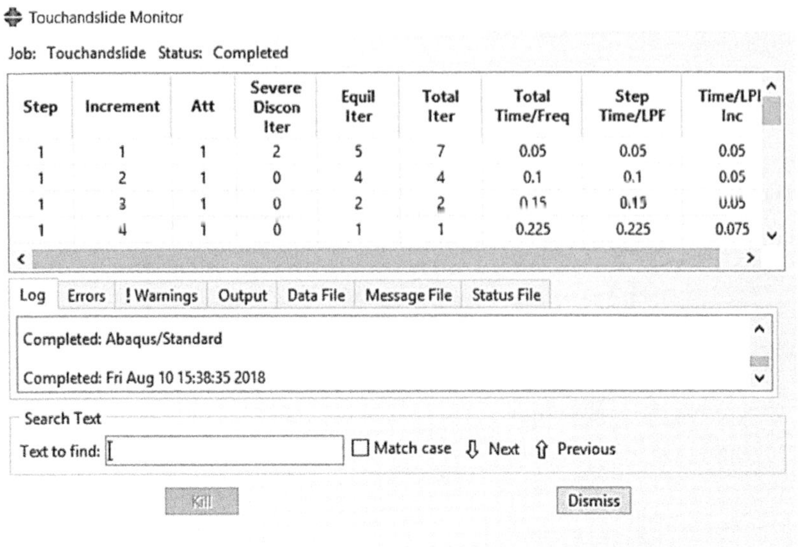

FIGURE 4.48
Monitoring the incrementation process. © Dassault Systèmes, a French "société européenne" (Versailles Commercial Register # B 322 306 440), or its subsidiaries in the U.S. and/or other countries.

of the model and the changes in descriptions throughout the deformation as the animation progresses. This step supports the final interpretation of the results and the drawing of conclusions about the model. A summary of changes in the parameters can be found in the *Report/Field* output (in the task bar). *Plotting Graphs* can be found under the *Tools* tab.

The following set of images (Figure 4.49) illustrates the distribution of Von Mises stresses on the skin model, depending on the stage of skin and yarn deformation. The model shows a cylinder representing a piece of textile (a yarn or a fiber) touching (i.e. exerting an effect by indenting) this skin. It takes 1 s to complete the touch phase. In the next phase, sliding, the cylinder slides over the surface of the skin, imitating typical human activity when assessing a fabric, called hand evaluation. The sliding phase also takes 1 s.

4.2.4 Changing the Model and Using Different Visualizations

The user can or even should introduce changes to the model to improve and correct it. Manipulate with parameters of the model, so you actually perform some modeling. The following outcome is the result of introducing certain changes to the original model, as described in Steps 1–12. The major difference here is that the load plate is retained in the visualization. In the same way as in Figure 4.49, Figure 4.50 shows a set of images illustrating the phases of major skin deformation due to the touch and sliding of the cylinder.

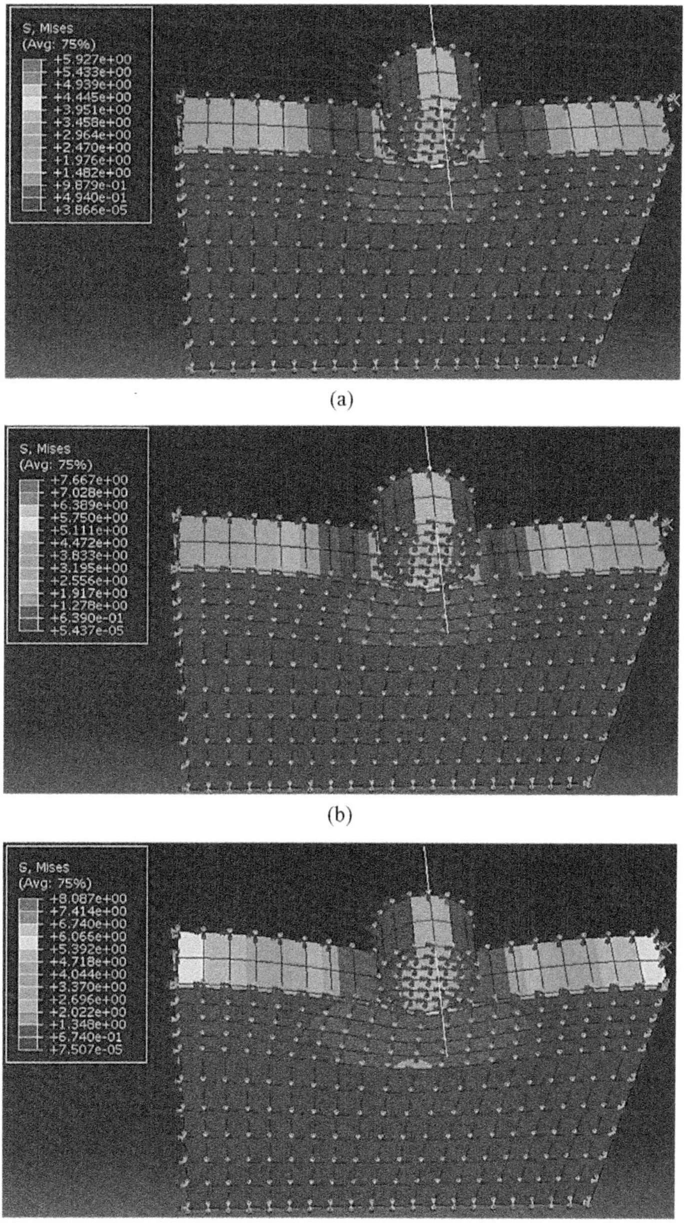

FIGURE 4.49
Von Mises stress (S) shown for a complete model composed of two parts (skin and yarn) as the skin deforms (indents) under the influence of the yarn while simulating the process of touching: skin deformation at the initial stage of touch (a) at 0.2 s of the touching phase, (b) 0.4 s, and (c) 0.6 s. © Dassault Systèmes, a French "société européenne" (Versailles Commercial Register # B 322 306 440), or its subsidiaries in the U.S. and/or other countries.

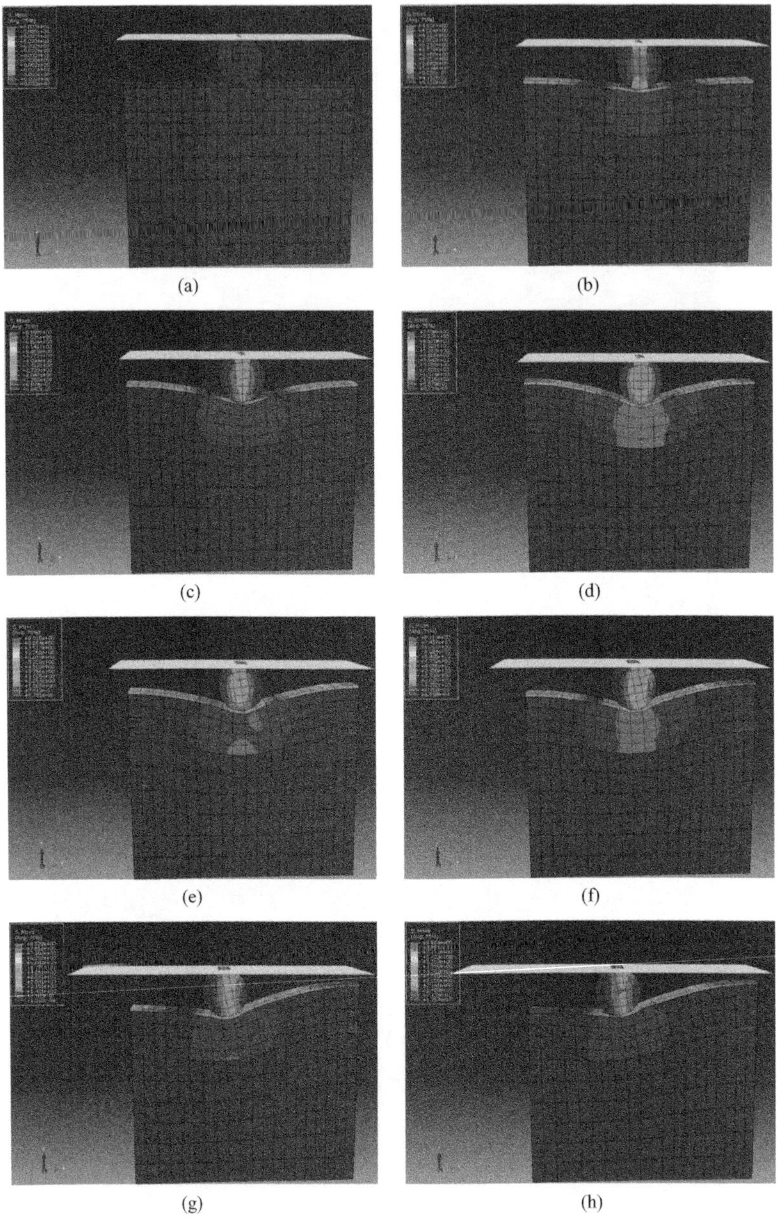

FIGURE 4.50
Von Mises stress (S) shown for a complete model composed of three parts (the skin, the yarn, and the load plate) as the skin deforms (indents) under the influence of the yarn in the simulated processes of touching and sliding on the surface of the skin: (a) undeformed skin at 0 s, (b) touching at 0.2 s, (c) 0.5 s, and (d) 1 s; touching and sliding at (e) 0.2 s, (f) 0.4 s, (g) 0.7 s, and (h) 1 s. © Dassault Systèmes, a French "société européenne" (Versailles Commercial Register # B 322 306 440), or its subsidiaries in the U.S. and/or other countries.

References

Barker, R.L. (2002). From fabric hand to thermal comfort: The evolving role of objective measurements in explaining human comfort response to textiles. *Int. J. Cloth Sci. Technol.* 14: 181–200.

Beech, S.R., et al. (1988). *Textile Terms and Definitions*, 8th edn. Textile Institute, Manchester.

Behery, H.M. (2005). *Effects of Mechanical and Physical Properties on Fabric Hand*, 1st edn. Woodhead Publishing Limited: Cambridge.

Ciesielska-Wróbel, I. and Van Langenhove, L. (2012). The hand of textiles - definitions, achievements, perspectives - a review. *Tex. Res J.* 82(14): 1457–1468.

Ciesielska-Wrobel, I.L. (2014). Fingertip skin models for analysis of the haptic perception of textiles. *J. Biomed. Sci. Eng.* 7: 1–6.

Dandekar, K., Raju, B.I., and Srinivasan, M.A. (2003). 3-D finite-element models of human and monkey fingertips to investigate the mechanics of tactile sense. *J. Biomech. Eng.* 125: 682–691.

Dassault Systèmes. (2014). Abaqus/CAE User's Guide.

Delalleau, A., Josse, G., Lagarde, J., Zahouani, H., and Bergheau, J. (2006). Characterization of the mechanical properties of skin by inverse analysis combined with the indention test. *J. Biomech.* 39: 1603–1610.

Derler, S., Schrade, U. and Gerhardt L.-C. (2007). Tribology of human skin and mechanical skin equivalents in contact with textiles. *Wear* 263: 1112–1116.

Fabric Assurance by Simple Testing at http://csiro.au/Portals/Publications/Brochures--Fact-Sheets.aspx. Accessed at 20 January 2016.

Fabric Touch Tester's description at http://sdlatlas.com/product/478/FTT-Fabric-Touch-Tester. Accessed on 20th January 2016.

Flynn, C.T. (2011). Mechanical characterisation of in vivo human skin using a 3D force-sensitive micro-robot and finite element analysis. *Biomech. Model Mechanobiol.* 10: 27–38.

Gassan, J., Chate, A., and Bledzki, A. (2001). Calculation of elastic properties of natural fibers. *J. Mater. Sci.* 36: 3715–3720.

Gibson, V.L. and Postle, R. (1978). An analysis of the bending and shear properties of woven, double-knitted, and warp knitted outwear fabrics. *Textile Res J.* 48: 14–27.

Groves, R.B., Coulman, S.A., Birchall, J.C., & Evans, S.L. (2013). An anisotropic, hyperelastic model for skin: Experimental measurements, finite element modelling and identification of parameters for human and murine skin. *J. Mech. Behav. Biomed. Mater.* 18: 167–180.

Hallos, R.S., Burnip, M.S., Weir, A. (1990). The handle of double-jersey knitted fabrics, Part I: Polar profiles. *J. Text. Inst.* 81: 15–35.

Handle-O-Meter description at http://thwingalbert.com/handle-o-meter.html. Accessed on 20 January 2016.

Hennes & Mauritz Group annual report for 2011 at http://about.hm.com/content/dam/hm/about/documents/en/Annual%20Report/Annual_Report_2011_P2_en.pdf. Accessed at 28 January 2016.

Hennes & Mauritz Group annual report for 2014 at http://about.hm.com/content/dam/hm/about/docume.nts/en/Annual%20Report/Annual%20Report%20 2014_en.pdf. Accessed at 28 January 2016

KES system for HoT measurements at http://english.keskato.co.jp/products/kes_fb4.html. Accessed on 20 January 2016.

Lapeer, R.J., Gasson, P.D., and Karri, V. (2011). A hyperelastic finite-element model of human skin for interactive real-time surgical simulation. *IEEE Trans. Biomed. Eng.* 58(4): 1013–1022.

Lumpkin, E.A. and Caterina, M.J. (2007). Mechanisms of sensory transduction in the skin. *Nature*, 445, 22.

Mark & Spencer brand multi-channel purchase information: Report and Financial Statements of 2011 at http://corporate.marksandspencer.com/documents/publications/2011/annual%20report%202011. Accessed on 20th January 2016.

Meilgaard, M.C., Carr, B.T. and Civille, G.V. (2007). *Sensory Evaluation Techniques,* 4th edn. CRC Press, Taylor and Francis Group, Boca Raton, FL.

New Look brand online shopping information: "New Look's 2011 Annual Report and Accounts" at http://apax.com/media/159679/New%20Look%20Annual%20Report%202010-11.pdf.pdf. Accessed on 20 January 2016.

On-line customers' debate at http://debate.org/opinions/is-it-still-important-to-consumers-to-be-able-to-touch-and-feel-products-in-a-physical-store. Accessed on 20 January 2016.

Park, S.W., Hwang, Y.G., Kang, B.C., et al. (2001). Total handle evaluation from selected mechanical properties of knitted fabrics using neural network. *Int. J. Cloth Sci. Technol.* 13: 106–114.

Peirce, F.T. (1930). The handle of cloth as a measurable quality. *J. Text. Inst.* 21: 337–416.

Penava, Ž., Šimić Penava, D., and Knezić, Ž. (2014). Determination of the elastic constants of plain woven fabrics by a tensile test in various directions. *FIBRES & TEXTILES Eastern Europe* 22(2(104)): 57–63.

PhabrOmeter's description at http://phabrometer.com/FAQ/pgePhabrOmeter.aspx. Accessed on 20 January 2016.

Pfarr, L. and Zagar, B. (2017). In-vivo human skin to textiles friction measurements. *17th Autex Conference.* IOP Conference Series: Materials Science and Engineering, Corfu, Greece, pp. 1–6.

Radhakrishnaiah, P., Tejatanalert, S, and Sawhney A. (1993). Handle and comfort properties of woven fabrics made from random blend and cotton – covered cotton/polyester yarns. *Text. Res. J.* 63: 573–579.

Ramamurthy, P., Chellamani, K., Dhurai, B., & Subramaniam, V. (2017). Study on frictional characteristics of medical wipes in contact with mechanical skin equivalents. *FIBRES & TEXTILES East. Eur.* 25(2(12)): 120–127.

Ramkumar, S.S., Wood, D.J., Fox, K. and Harlock, S.C. (2003a). Developing a polymeric human finger sensor to study the frictional properties of textiles. Part I: Artificial finger development. *Text. Res. J.* 73(6): 469–473.

Ramkumar, S.S., Wood, D.J., Fox, K. and Harlock, S.C. (2003b). Developing a polymeric human finger sensor to study the frictional properties of textiles: Part II: Experimental results. *Text. Res. J.* 73(7): 606–610.

Srinivas, K., Lakshumu Naidu, A., and Raju Bahubalendruni, M. (2017). A review on chemical and mechanical properties of natural fiber reinforced. *Int. J. Perform. Eng.* 13(2): 189–200.

Stearn, A.E., D'Arcy, R.L., et al. (1988). A statistical analysis of subjective and objective methods of evaluating fabric handle, Part I: Analysis of subjective assessments. *J Text. Mach. Soc. Jpn.* 34: 13–18.

Tada, M., Nagai, N., Yoshida, H. and Maeno, T. (2006). Iterative FE analysis for non-invasive material modelling of a fingertip with layered structure. *Proceedings of the European Haptics Conference,* 2006.

Teijin (2018). *Mechanical properties of fibers.* Retrieved October 4, 2018, from https://teijin.com/products/advanced_fibers/aramid/contents/aramid/conex/eng/bussei/conex_bussei_hippari.htm.

Tissue Softness Analyzer's description at http://emtec-papertest.de/en/tsa-tissue-softness-analyzer.html. Accessed on 20 January 2016.

Vilhena, L. and Ramalho, A. (2016). Friction of human skin against different fabrics for medical use. *Lubricants* 4(6): 1–10.

Xu, F., Wen, T., Lu, T., and Seffen, K. (2008). Skin biothermomechanics for medical treatments. *J. Mech. Behav. Biomed. Mater.* 1(2): 172–187.

Yokura, H. and Niwa, M. (2003). Objective hand measurement of nonwoven fabrics used for the top sheets of disposable diapers. *Tex. Res J.* 73: 705–712.

Zara annual reports for 2013 at http://inditex.com/documents/10279/18789/Inditex_Annual_Report_2014_web.pdf/a8323597-3932-4357-9f36-6458f55ac099. Accessed at 20 January 2016.

Zara annual reports for 2014 at http://inditex.com/documents/10279/18789/Inditex_Group_Annual_Report_2013.pdf/88b623b8-b6b0-4d38-b45e-45822932ff72. Accessed at 20 January 2016.

5

Geometry of a Knit Structure

5.1 Knit Structure

Knitting is one of the most popular textile technologies, and is used worldwide for many purposes, predominantly for all sorts of clothing; however, applications of knit structures occur in technical textiles (e.g., screening against electromagnetic radiation, auxetic structures against impact, and shoes).

We distinguish between two basic knitting classes: weft and warp. This division is related to the fact that the basic elements composing these two groups of stitches are weft loops and warp loops. There are some typical characteristic elements of the weft and warp types of loops, as presented in Figure 5.1.

The weft loop is composed of a top arc (the so-called needle arc), arc 34; two bottom half arcs, arc 12 and arc 56; elements connecting the bottom of the loop with the top are the loop legs: $\overline{23}$ and $\overline{45}$.

The warp loop is composed of a top arc, arc 23; elements connecting the bottom of the loop with the top are loop legs: $\overline{12}$ and $\overline{34}$; and the bottom arc, arc 45.

Figure 5.1 presents a schematic two-dimensional (2D) view of the loops in knitting structures. It is possible to model a 2D knit structure in Abaqus software. However, it is believed that three-dimensional (3D) visualization and

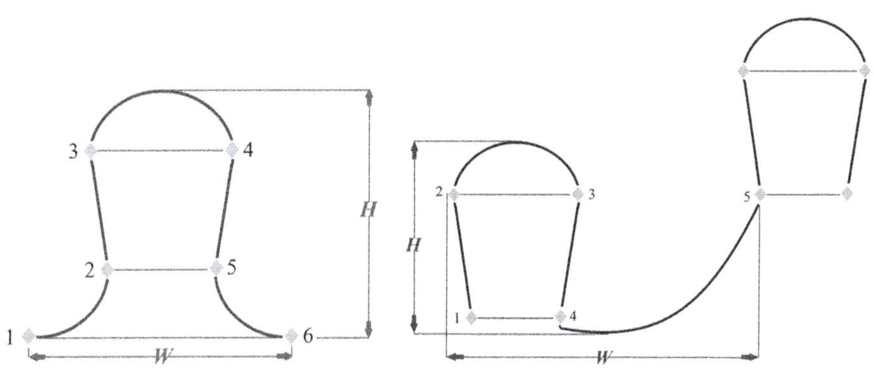

FIGURE 5.1
Basic types of loops in knitting: left—weft loop, right—warp loop; *W*—the width of the loop; *H*—the height of the loop.

structure simulation is more spectacular. This chapter will provide the basic information allowing the modeling of one of the basic knit textile structure, the purl stitch, and its basic element, the loop.

A single loop or a row might be quite challenging to draw and create a geometry suitable for further modeling in Abaqus. Therefore, some measures will be taken to make this drawing process successful and as a consequence able to operate with the draw of the loop.

It is possible to draw a single loop or a row using SOLIDWORKS® or other 3D design environments and subsequently import them to Abaqus. It is also possible, and this method will be discussed here, to create the geometry of a single loop and, based on it, to create a row using Abaqus.

There are two different sets of instructions presented here. The first one offers the steps for creating a simplified loop geometry. The second one requires more steps to be completed to create a more sophisticated geometry.

5.2 Guide for Sketching a Simplified Loop

As in all previous steps when creating other models, one starts from *Create Part.*

This time, one selects the following setting for the model of loop, being a basic element of the stitch: Modeling space—3D, Modeling type—Deformable, Solid body, and creation via sweep. After confirming the selected parameters by *Continue,* as presented in Figure 5.2, one moves to the actual sketching of the geometry. For some, sketching only half of the loop and combining both halves later on may be easier than drawing the whole loop in one step. This is especially true when introducing some partitions into the shape of the loop aiming at shaping it and having it be more sophisticated.

As always, after confirming the selection of initial parameters, the background grid appears and the user is ready to create a geometry. It is suggested to click on the grid in the central part of the screen to fix an arbitrary center. However, this tip is meant to help to organize drawings on the grid, and thus it is purely esthetic. It is suggested to use a tool called *Create Lines: Connected,* although any other means that can lead to a correctly created model is very welcome here. This launches the line and initiates sketching, as presented in Figure 5.3. After a final click, the desired line appears on the grid. Its dimensions can be fixed by the selection of a tool called *Add Dimension.* Having clicked on *Create Circle,* one can draw the circle and fix its radius.

In order to shape the curves, one should use *Created Fillet: Between 2 Curves.* This step requires providing the radius R for the curved shape, which in this case is 12.5 [mm].

Select the first and second entities near the end of the line to be filled and repeat the step for the second right angle. Observe the prompt area for the

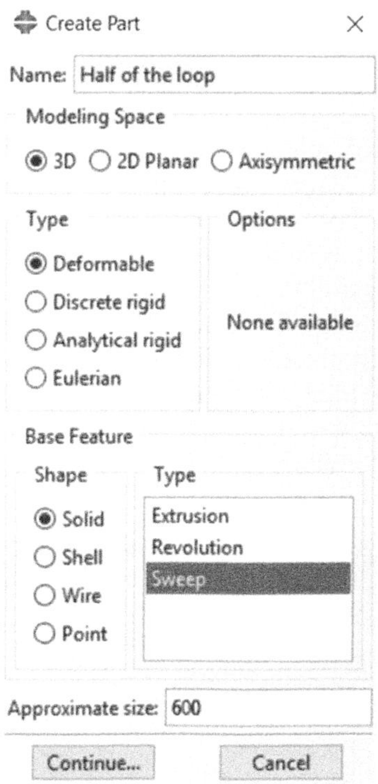

FIGURE 5.2
Create Part submenu allowing the creation of a geometry that is a 3D deformable solid body created in sweep mode. © Dassault Systèmes, a French "société européenne" (Versailles Commercial Register # B 322 306 440), or its subsidiaries in the U.S. and/or other countries.

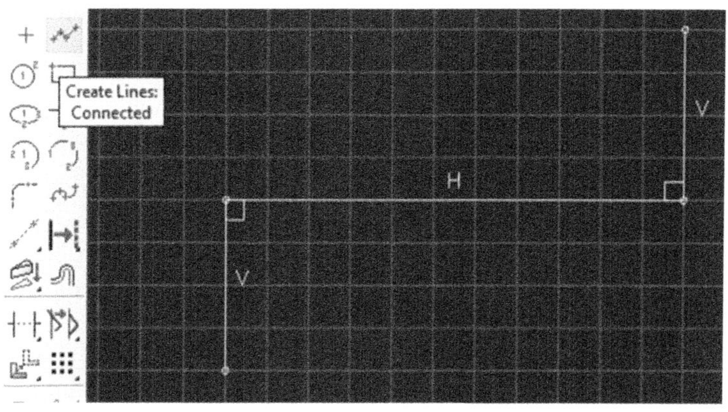

FIGURE 5.3
Selection of the tool *Create Circle* for creating a geometry. © Dassault Systèmes, a French "société européenne" (Versailles Commercial Register # B 322 306 440), or its subsidiaries in the U.S. and/or other countries.

tips, refrain from unnecessary steps or select the desired step if needed (e.g., if no further filled lines are needed, as presented in Figure 5.4).

The next tip appearing in the prompt area is Sketch the sweep path. It requires confirmation. One is also required to type in a scale for the section sketch. In the current example, the 600 scale has been used.

After this, the sketch plane sweeps around and gets centralized, so the sketch lines are barely visible and one looks at them top down. To make them more visible, it is advised to use a tool called *Rotate View* to obtain a view as presented in Figure 5.5.

At this point one needs to sketch a circular cross section that is going to be swept along the red path presented in Figure 5.6a and b.

To do so, one may use the *Create Circle: Circle and Perimeter* tool.

As soon as one establishes the parameters of the cross-section and confirms its choice, the software generates a 3D object being a first half of the loop, as shown in Figure 5.7.

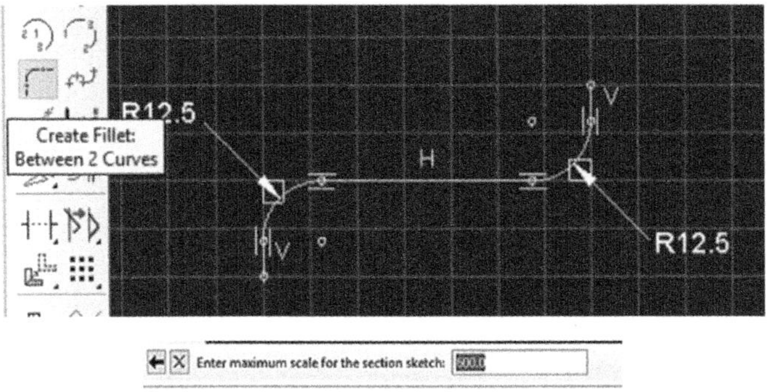

FIGURE 5.4
Creating fillets. © Dassault Systèmes, a French "société européenne" (Versailles Commercial Register # B 322 306 440), or its subsidiaries in the U.S. and/or other countries.

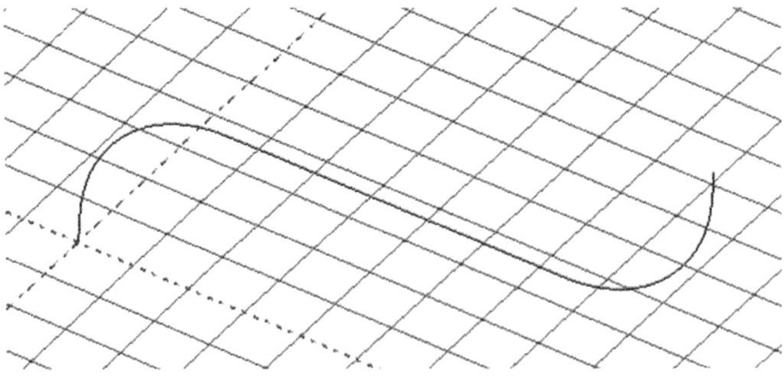

FIGURE 5.5
To help to observe the details, the regular dark blue background was changed to white (changing the background color is possible via View/Graphics Options/Viewport Background. © Dassault Systèmes, a French "société européenne" (Versailles Commercial Register # B 322 306 440), or its subsidiaries in the U.S. and/or other countries.

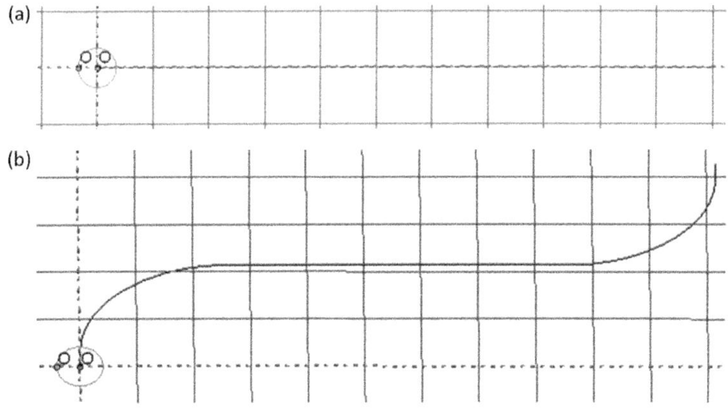

FIGURE 5.6
Sketching a circular cross section and sweeping it along a path: (a) a view from the top and (b) a side view of created sketch. © Dassault Systèmes, a French "société européenne" (Versailles Commercial Register # B 322 306 440), or its subsidiaries in the U.S. and/or other countries.

FIGURE 5.7
Created element of the loop. © Dassault Systèmes, a French "société européenne" (Versailles Commercial Register # B 322 306 440), or its subsidiaries in the U.S. and/or other countries.

Merging two halves of the loop is possible in *Assembly*. To do so, select Instances. It will automatically change the color of the part into blue and open a new window called *Create Instance*. In this case, one operates on the base of a part (half of the loop); thus, one selects a specific part (e.g., Part-1, in this *Create Instance* window). After confirmation of the selection, the new Instance appears on the list of instances in the *Model tree*.

Operating within the *Assembly* module allows using specific tools meant for multiplication of the instances. However, in this case, it is enough to repeat the last step, meaning the creation of a new instance, which will result in having another identical instance on the list of instances in the *Model tree*. It is easy to observe both instances listed in the *Model tree*; however, it is not always easy to see both on the screen, since usually after adding two identical instances, they both appear on the screen in exactly the same position. Thus, a user can see only one element, although both instances are present in the system. In order to create a full loop (i.e. merge both halves of the loop) and/or to see both halves on the screen, one may use one of the tools present on the left side of the *Assembly* module. In this case, it is suggested to select the *Rotate Instance* tool, which allows shifting one of the instances and placing it in the right position. To do so, one selects the instance to be rotated on the screen, as the comment in the prompt area suggests. After confirming the selection, one is asked to select a starting point for the axis of rotation or enter *X,Y,Z* coordinates.

While appropriate comments appear under the instance in the prompt area, the instance itself also adapts to allow introducing changes (e.g., the blue colored shape of the instance is outlined by a red line and the characteristic points of the instance are marked by yellow). These are the colors usually used. One selected a point on the edge of the tubular shape of the half of the loop as presented in Figure 5.8. Selection of the point changes its color.

After selecting the starting point, one is asked to select the end point for the axis of rotation. Here, a point lying on the other side of the cross-sectional circle has been selected. This action changes this second point's color and provides another tip presented in the prompt area. This time, it is a question about the angle of rotation of the instance. One is trying to build a full loop shape; thus, in this case, one selects 180° as the angle of rotation for the instance. The merged halves are presented in Figure 5.9, and the multiplied loop is presented in Figure 5.10. In order to multiply the loop, one used the Linear Pattern tool.

Naturally, such a simplified shape of the loop is very far from the reality and from the most typical shape of the loop used in different modeling practices.

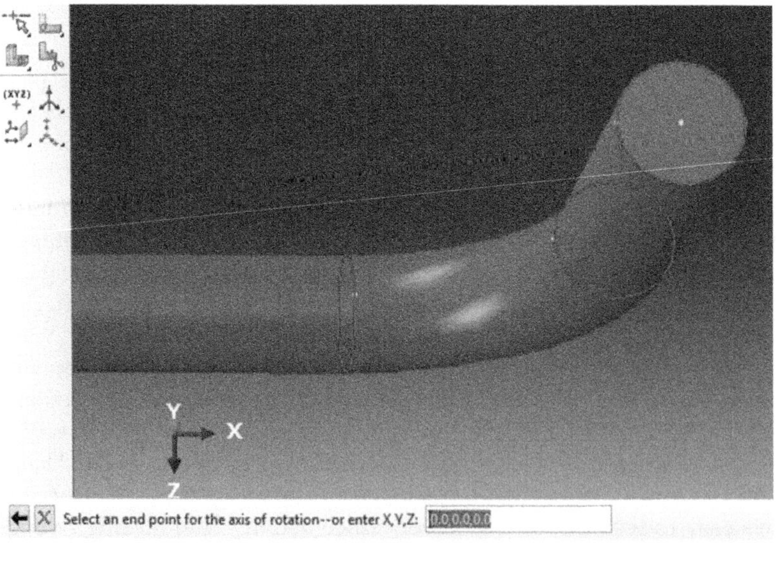

FIGURE 5.8
Selection of a point on the edge of a tubular shape. © Dassault Systèmes, a French "société européenne" (Versailles Commercial Register # B 322 306 440), or its subsidiaries in the U.S. and/or other countries.

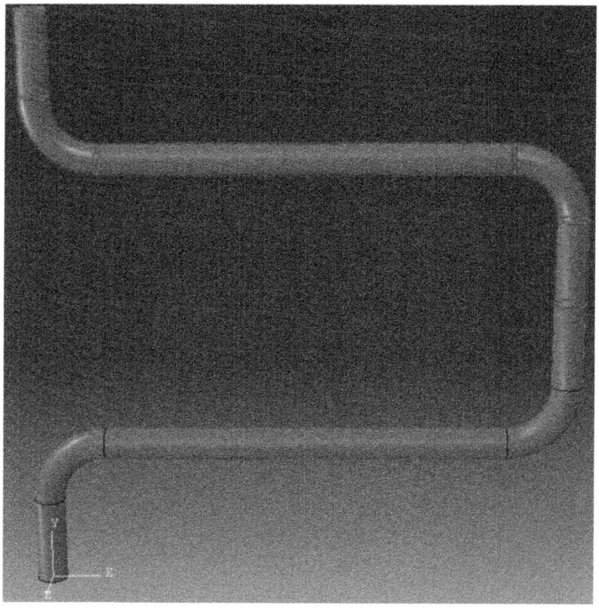

FIGURE 5.9
Two merged halves of the loop. © Dassault Systèmes, a French "société européenne" (Versailles Commercial Register # B 322 306 440), or its subsidiaries in the U.S. and/or other countries.

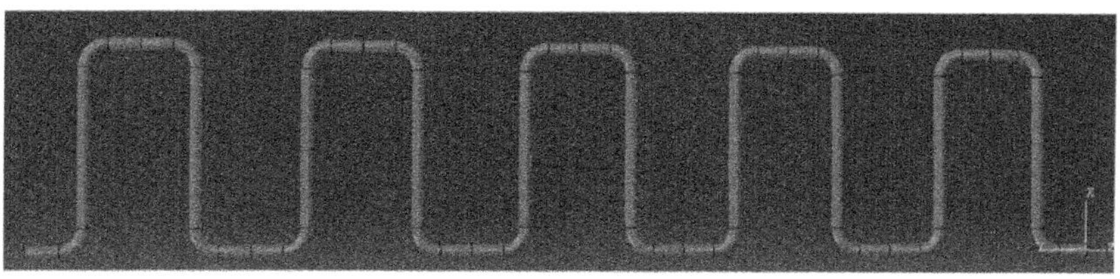

FIGURE 5.10
Multiplied loop. © Dassault Systèmes, a French "société européenne" (Versailles Commercial Register # B 322 306 440), or its subsidiaries in the U.S. and/or other countries.

To obtain a more realistic shape of the loop, it is advised to sketch elements connecting the bottom of the loop with its top (as presented in Figure 5.1: $\overline{23}$ and $\overline{45}$) at a specific angle. Thus, instead of the initial sketch as presented in Figure 5.3, Selection of the tool *Create Circle* for creation of the geometry, one sketches another type of loop, as suggested in Figure 5.11.

It does not matter whether the shape of the loop is more sophisticated or whether one sketches the whole shape at once or whether the halves of the loop are merged at a later stage. These loops may be easily connected in a row but cannot be easily connected in a knit structure, as the shape of the loop was not profiled adequately to make the connection in a whole structure possible and realistic. This is why it is crucial to sketch the loop in two planes.

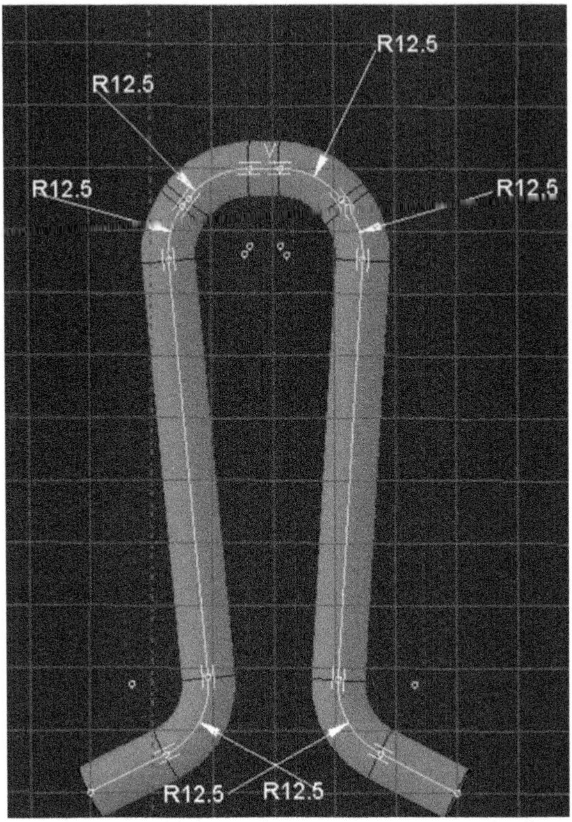

FIGURE 5.11
Sketching the shape of the loop.

5.3 Guide for Sketching a Profiled Loop

The three steps presented below describe in detail the creation of adequate geometry of a knit structure.

STEP 1: Creating the geometry of half of a single loop of a knit structure (Figure 5.12)

Idealized loops are symmetric. One uses here this symmetry property of the loop to simplify the process of creating the geometry and starts from drawing the first half of the loop. This is why a prior understanding of the architecture of a knit structure and its basic construction elements is required.

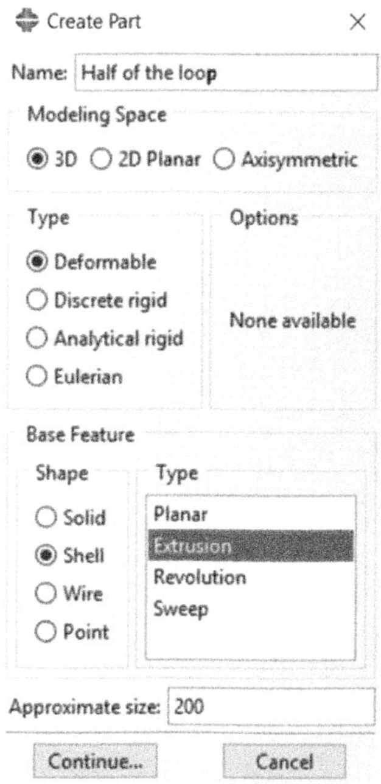

FIGURE 5.12
Create Part submenu allowing the creation of a geometry. © Dassault Systèmes, a French "société européenne" (Versailles Commercial Register # B 322 306 440), or its subsidiaries in the U.S. and/or other countries.

In the *model tree*, double click on *Parts* and select *Create*.

The name given to this model is *Half of the loop*.

This element is *3D, Deformable, Shell* body.

The way the model is to be created is *Extrusion*, so the corresponding selection should be made.

Approximate size is 200.

Sketching the shape of the loop is very similar to the process presented in Figure 5.2.

Any deviations and reshaping of the loop are possible, as presented in Figure 5.13.

After finalizing the sketch, a comment concerning extrusion appears in the prompt area.

Confirming the selection by the *Done* key evokes a new window called *Edit Feature*, where one may select the depth in mm, this being the third dimension of the model, as presented in Figure 5.14.

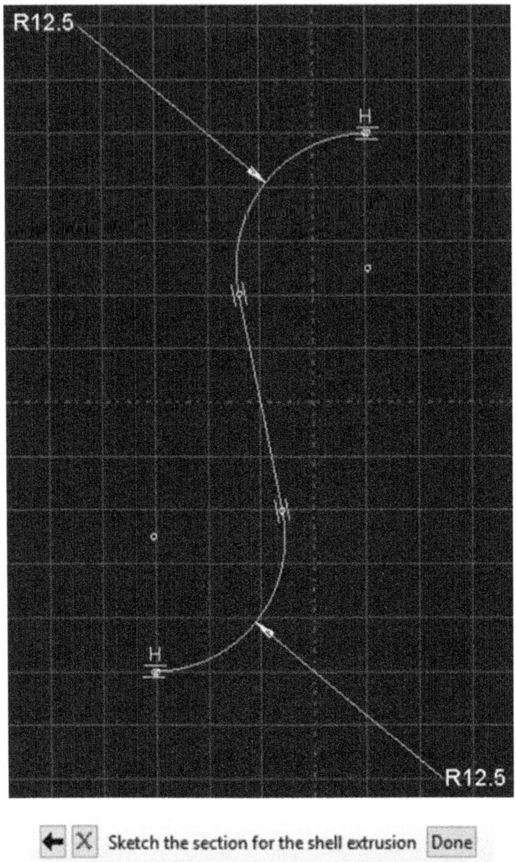

FIGURE 5.13
Example of a sketch of half of a loop together with a comment, which appears in the prompt area. © Dassault Systèmes, a French "société européenne" (Versailles Commercial Register # B 322 306 440), or its subsidiaries in the U.S. and/or other countries.

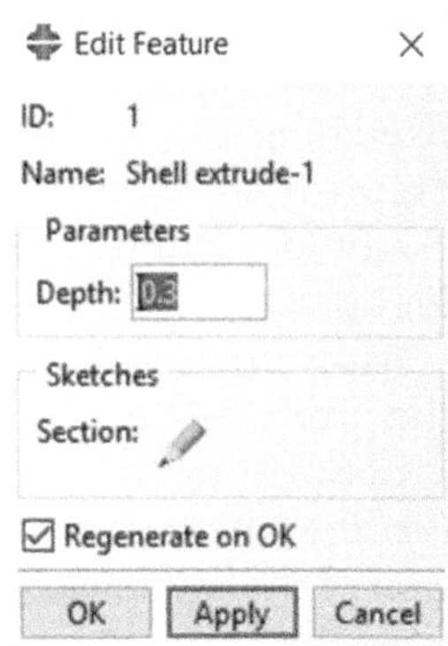

FIGURE 5.14
Higher values of depth allow easier manipulation with a shell and its edges. The value 3 is sufficient; the value 0.3 requires working on a magnified sketch. © Dassault Systèmes, a French "société européenne" (Versailles Commercial Register # B 322 306 440), or its subsidiaries in the U.S. and/or other countries.

STEP 2: Partition of the faces

Partition allows dividing the object, the shell in this case. After removing some redundant elements that are marked in a partition step, the shape that remains will be a more refined, carved shape of the final 3D model. This is why the shape of the shell needs to be partitioned a few times to reshape it and prepare a model that is more realistic, better reflects the loop, and allows its being combined into rows. The partitioned and refined shape provides a route/path for the final model to sweep along.

While still being in the Part module, select the partition tools, and specifically, one uses Partition Face: use the Curved Path Normal to 2 Edges tool, as presented in Figure 5.15.

The next step, which is creating the wire based on the sketch, is presented in Figure 5.16.

The creation of (c) from Figure 5.17 is preceded by a comment that all associated faces and cells will be deleted from the model so that the wire frame will be the only element remaining.

STEP 3: Create Datum axis and Datum plane

Both the Datum axis and Datum plane are additional geometry elements that are used as a support in the creation of a model. They do not have an actual impact on the model itself, and they do not undergo a meshing process. The only reason they come into play during modeling is to create additional lines or planes that are supportive when constructing a complex geometry of an element that will be part of an actual model.

To create a Datum axis, select the Datum Axis: Principal Axis tool as presented in Figure 5.18.

After choosing the Principal Axis tool, the prompt area displays a choice of three available axes. It is possible to use all of them; however, to continue

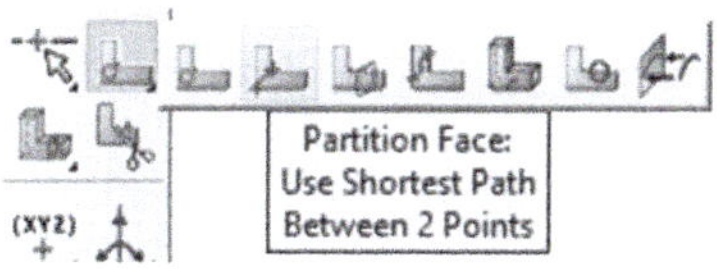

FIGURE 5.15
Partitioning the object. © Dassault Systèmes, a French "société européenne" (Versailles Commercial Register # B 322 306 440), or its subsidiaries in the U.S. and/or other countries.

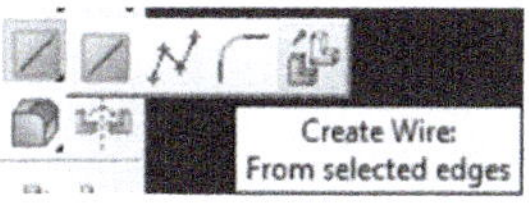

FIGURE 5.16
Creating the wire based on the sketch. © Dassault Systèmes, a French "société européenne" (Versailles Commercial Register # B 322 306 440), or its subsidiaries in the U.S. and/or other countries.

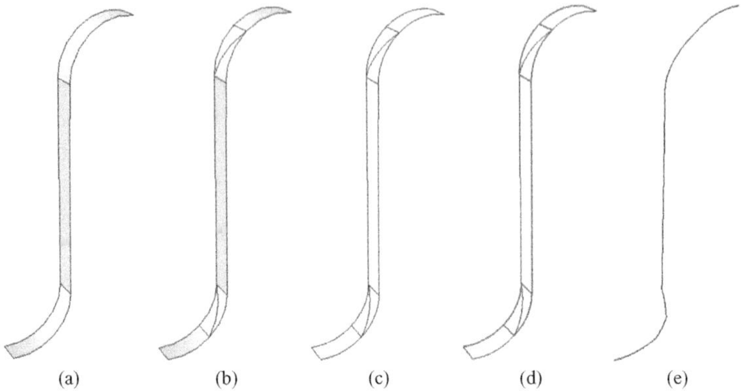

FIGURE 5.17
Stages of transforming the sketch into a wire using the tools presented in Figures 5.15 and 5.16: (a) original sketch, (b) a wire in the original sketch; (c) removing faces and cells from the sketch, (d) selecting the desired shape of the loop, and (e) removing redundant elements. © Dassault Systèmes, a French "société européenne" (Versailles Commercial Register # B 322 306 440), or its subsidiaries in the U.S. and/or other countries.

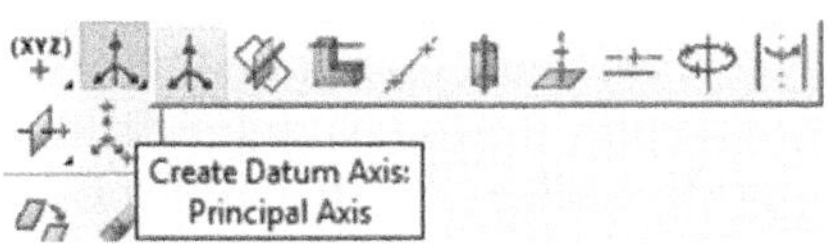

FIGURE 5.18
Set of available Datum Axis tools. © Dassault Systèmes, a French "société européenne" (Versailles Commercial Register # B 322 306 440), or its subsidiaries in the U.S. and/or other countries.

creating the current model, only one axis is needed, the x-axis. After selection of this axis, the appropriate geometrical element, the axis, appears as a dotted line in the vicinity of the created geometry of the half of the loop.

In the next step, one creates a Datum plane. There are a few methods to create a plane. However, the most suitable in this case is creating a Datum Plane by selecting a point and a normal. The tool for this can be found among the Datum Plane tools, as presented in Figure 5.19.

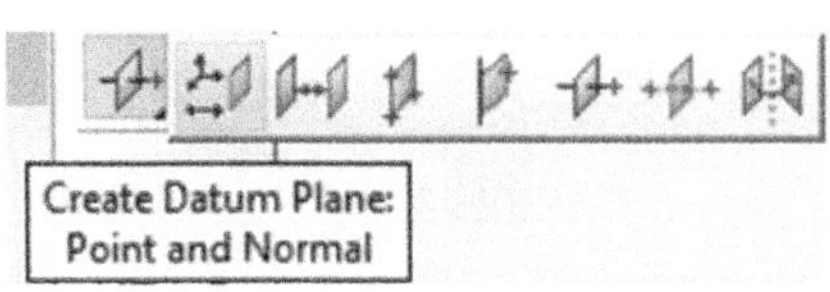

FIGURE 5.19
Set of available Datum Plane tools. © Dassault Systèmes, a French "société européenne" (Versailles Commercial Register # B 322 306 440), or its subsidiaries in the U.S. and/or other countries.

After selecting the appropriate tool, the prompt area indicates that it is required to select a point through which the datum plane will pass. One selects a tip of the wire as this required point. After this selection, a new requirement appears in the prompt area. This time one is asked to select a straight line normal to the datum plane. The normal to the datum plane in this case is the datum axis created earlier, and it should be selected now. Both datum elements are presented in Figure 5.20.

At this point, one should go back to the Create wire set of tools and select Create Wire: Planar. This means that the wire may be subject to further modifications in respect of the planar, as one of its tips has its beginning in the previously created planar.

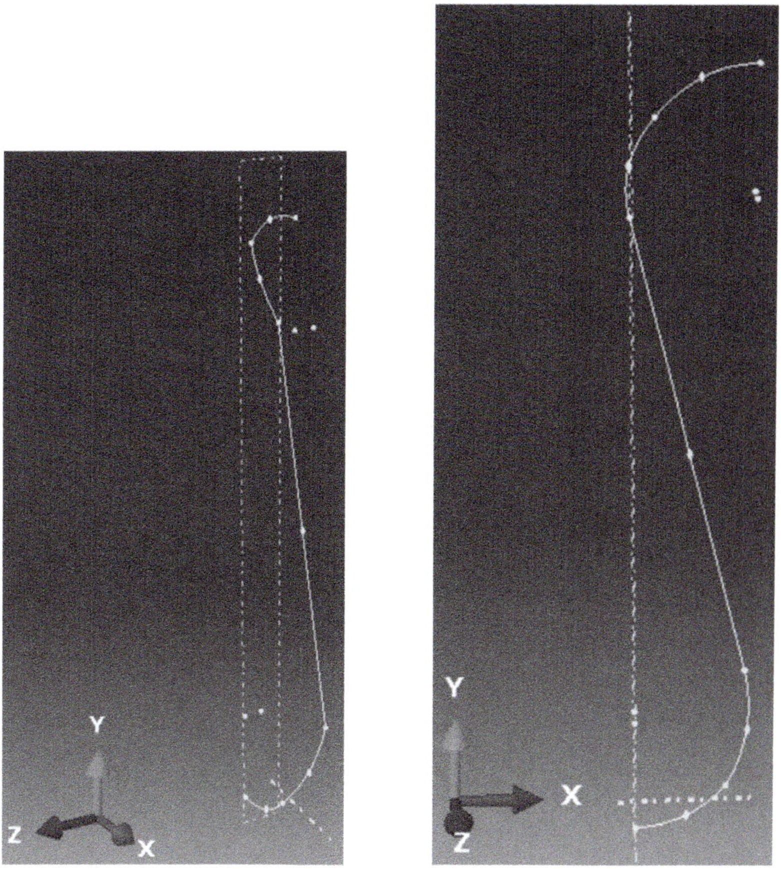

FIGURE 5.20
Different views of a wire, datum plane, and datum axis (at the bottom of the wire). © Dassault Systèmes, a French "société européenne" (Versailles Commercial Register # B 322 306 440), or its subsidiaries in the U.S. and/or other countries.

After selection of Create Wire: Planar, one is asked to select a plane for the planar wire. There is only one element of the geometry that could be selected at this point: it is the previously created Datum plane. As soon as it is selected, it changes color, and the prompt area displays another command. This time, one is asked to select an edge or an axis. One should select an edge that is a part of the wire, the one that has a direct contact with the plane. After selection of the part of the wire that has its tip lying in the plane, the viewport with the model changes into a sketching grid, as presented in Figure 5.21.

This is the moment when one may sketch an additional element, here a circle, indicating the shape of the cross section of the final model.

One selects Create Circle: Center and Perimeter from the sketching tools and draws a circle that has its center covering the tip of the wire. After completing the drawing, the prompt area requires confirmation by selecting the *Done* key.

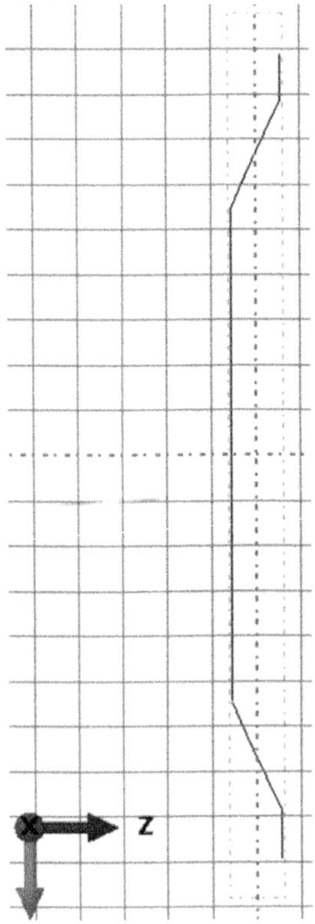

FIGURE 5.21
View of a sketching grid with a wire (black line) lying on the datum plane (a yellow dotted line). © Dassault Systèmes, a French "société européenne" (Versailles Commercial Register # B 322 306 440), or its subsidiaries in the U.S. and/or other countries

As a consequence, one receives the wire with a circle attached to one of its tips, as presented in Figure 5.22.

In the next step, one selects a tool called Create Shell: Sweep in order to sweep/cover a wire by a tubular shape. A consequence of this selection is opening a new window called Create Shell Sweep, as presented in Figure 5.23.

Editing the path along which the shell should appear only requires selecting the wire along which the shell should be swept, thus one should select one of the wire's edges and the direction of the sweep, which is clearly indicated on the wire in the viewport. After confirmation of the selections, one selects editing the profile's edge, and this time a prompt area displays a command saying that the connected set of edges should be selected. One selects the already created geometry of a circle. This step finishes the creation of a tubular shell around the wire, as it appears in the viewport.

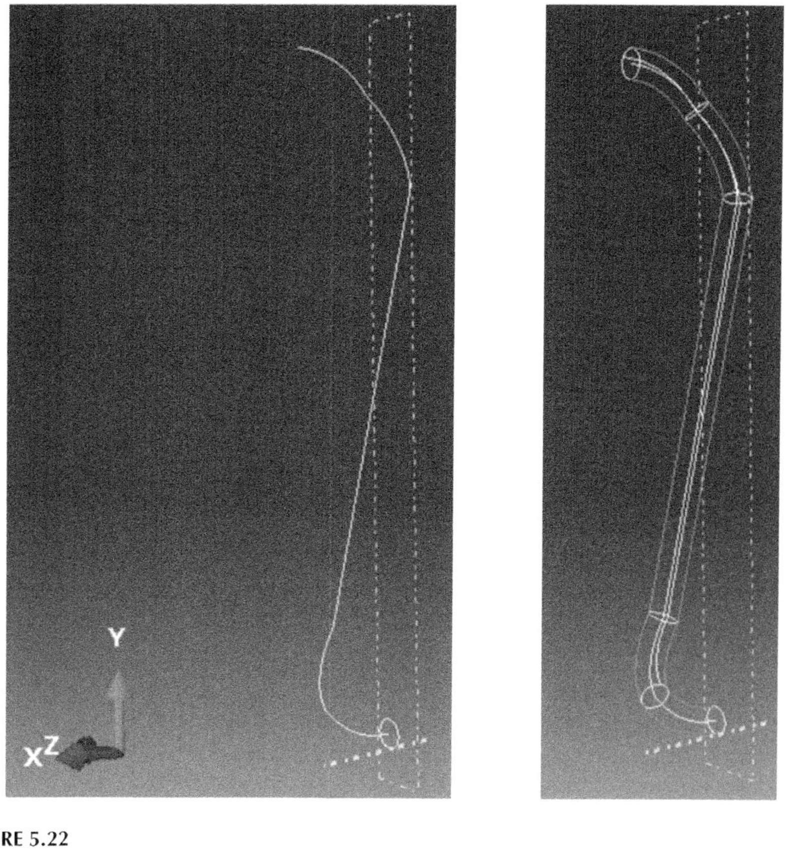

FIGURE 5.22
Stages of creating a tubular shape imitating a half of the loop. © Dassault Systèmes, a French "société européenne" (Versailles Commercial Register # B 322 306 440), or its subsidiaries in the U.S. and/or other countries.

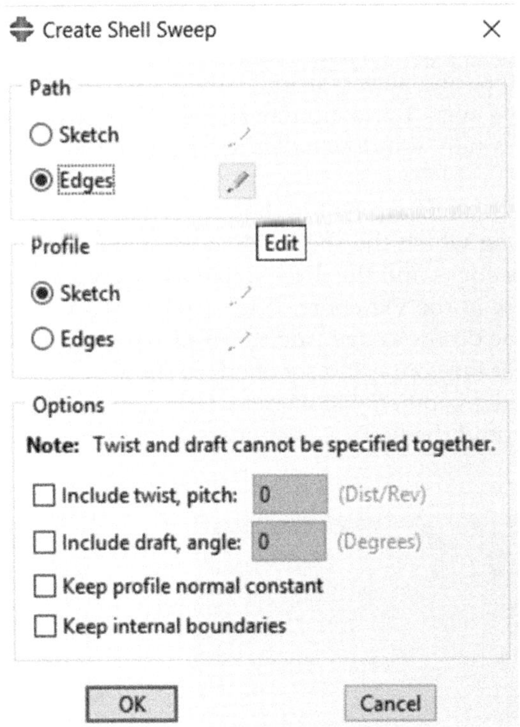

FIGURE 5.23
Create Shell Sweep window where one may edit both the path and the profile of the swept shell object. © Dassault Systèmes, a French "société européenne" (Versailles Commercial Register # B 322 306 440), or its subsidiaries in the U.S. and/or other countries.

In the next steps, one should remove the wire that is inside the tubular shape playing the role of half of the loop, and next close both outlets in this tubular shape.

To do so, one selects Geometry Edit tool. It evokes the appearance of Geometry Edit window in the viewport as presented in Figure 5.24.

After selecting Remove wire commend from the Geometry Edit window, one is asked to select wires, which in this case is performed individually. It is enough to select the wire, and its selection changes its color to red.

In the next step, one closes the tubular outlets by covering them. To complete this step, one selects Cover Edges tool. The prompt area displays an appropriate tip that suggests selection of the edge that should be covered. This operation should be performed for both outlets of the tubular shell, as presented in Figure 5.25.

In the next steps, one should follow all the typical steps allowing the creation of the model. One of the important steps in the case of a half of the loop is meshing it and merging it with another half, see Figure 5.26.

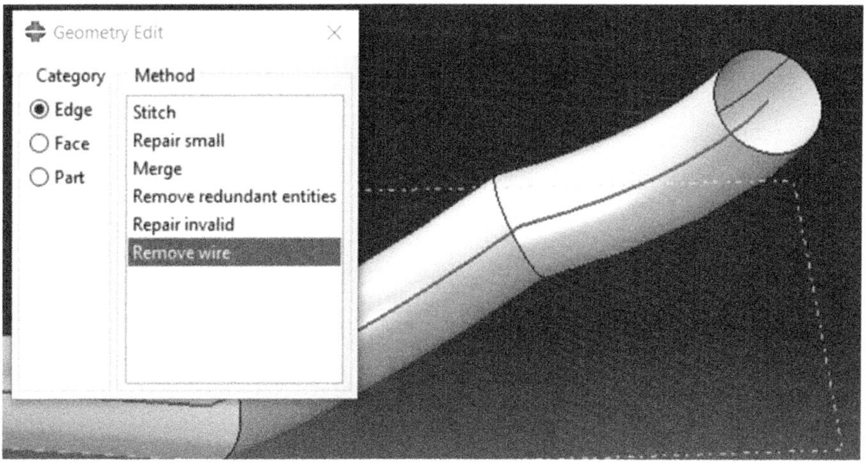

FIGURE 5.24
Removing the wire from the inside of a tubular shell by using Geometry Edit/Remove wire tool. © Dassault Systèmes, a French "société européenne" (Versailles Commercial Register # B 322 306 440), or its subsidiaries in the U.S. and/or other countries.

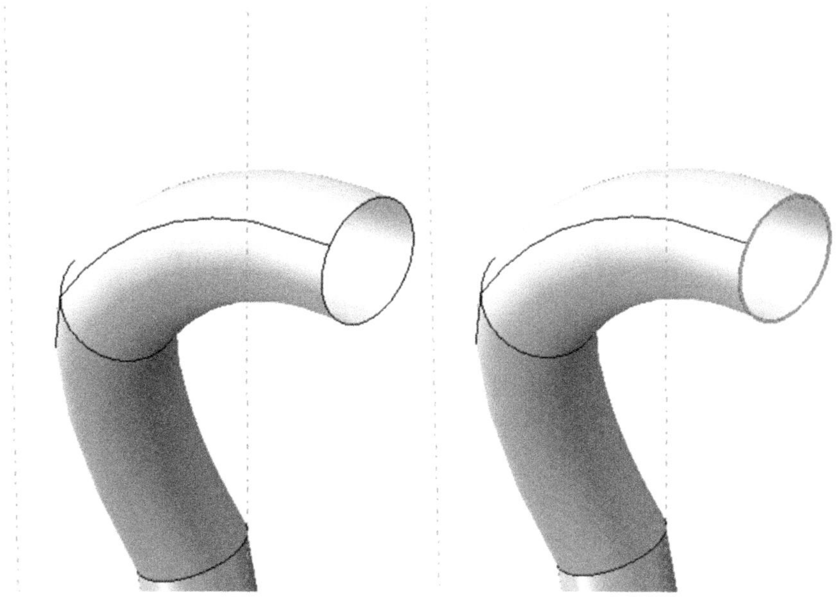

FIGURE 5.25
Using Cover Edges tool when closing the tubular outlet: (a) open outlet, (b) marked edge of the outlet, and (c) covered outlet. © Dassault Systèmes, a French "société européenne" (Versailles Commercial Register # B 322 306 440), or its subsidiaries in the U.S. and/or other countries.

(Continued)

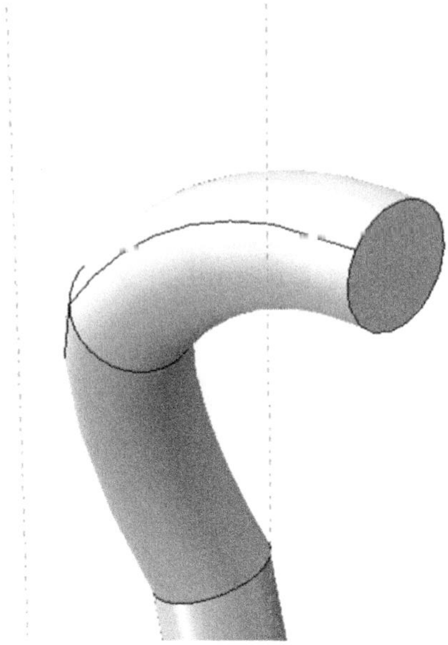

FIGURE 5.25 (*Continued*)
Using Cover Edges tool when closing the tubular outlet: (c) covered outlet. © Dassault Systèmes, a French "société européenne" (Versailles Commercial Register # B 322 306 440), or its subsidiaries in the U.S. and/or other countries.

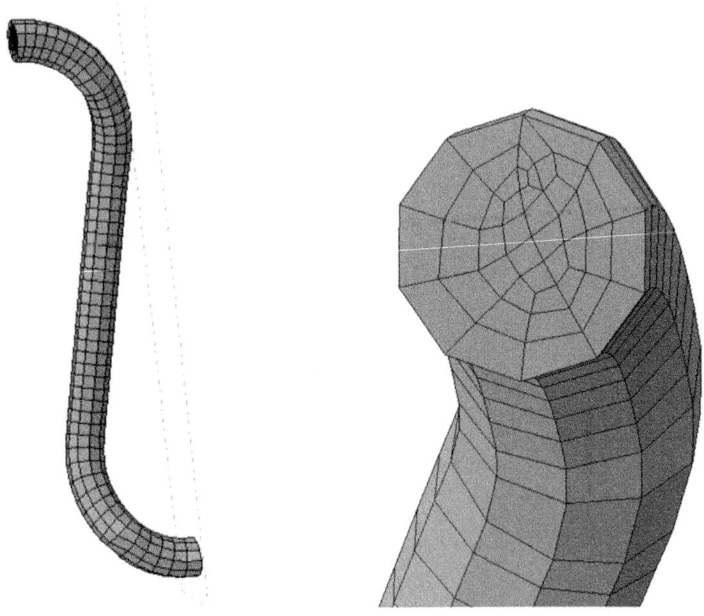

FIGURE 5.26
Meshed half of the loop. © Dassault Systèmes, a French "société européenne" (Versailles Commercial Register # B 322 306 440), or its subsidiaries in the U.S. and/or other countries.

Further smoothing the mesh and merging the halves allows obtaining the models presented in Figure 5.27. A half of the loop contains 2,365 elements after meshing.

There are endless possibilities for using the models of a knit structure. An example is presented in Figure 5.28, where a knit structure is touched by the skin section.

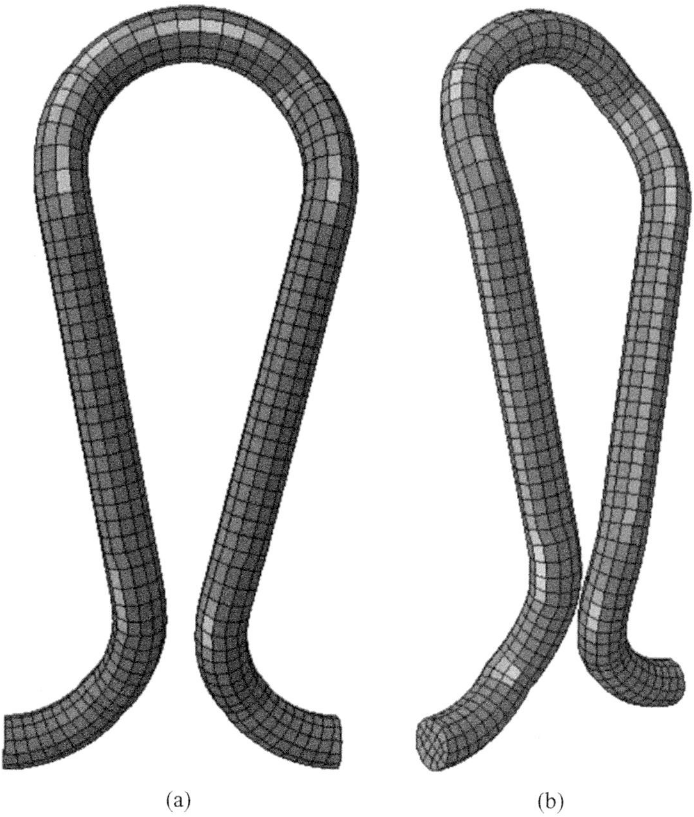

(a) (b)

FIGURE 5.27
(a) and (b) Views of a single loop, (c) view of row, (d) a model of a purl stitch's fabric face, (e) side view of (d), and (f) magnified connection points of the yarns of two different rows. © Dassault Systèmes, a French "société européenne" (Versailles Commercial Register # B 322 306 440), or its subsidiaries in the U.S. and/or other countries.

(Continued)

(c)

(d)

FIGURE 5.27 (*Continued*)
(a) and (b) Views of a single loop, (c) view of row, (d) a model of a purl stitch's fabric face, (e) side view of (d), and (f) magnified connection points of the yarns of two different rows. © Dassault Systèmes, a French "société européenne" (Versailles Commercial Register # B 322 306 440), or its subsidiaries in the U.S. and/or other countries.

(Continued)

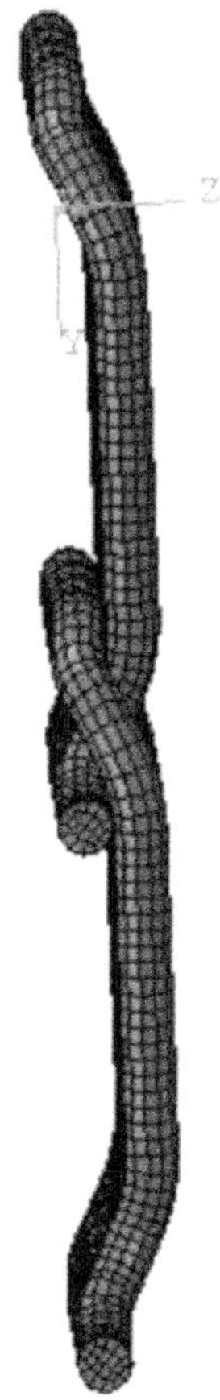

(e)

(f)

FIGURE 5.27 ***(Continued)***
(a) and (b) Views of a single loop, (c) view of row, (d) a model of a purl stitch's fabric face, (e) side view of (d), and (f) magnified connection points of the yarns of two different rows. © Dassault Systèmes, a French "société européenne" (Versailles Commercial Register # B 322 306 440), or its subsidiaries in the U.S. and/or other countries.

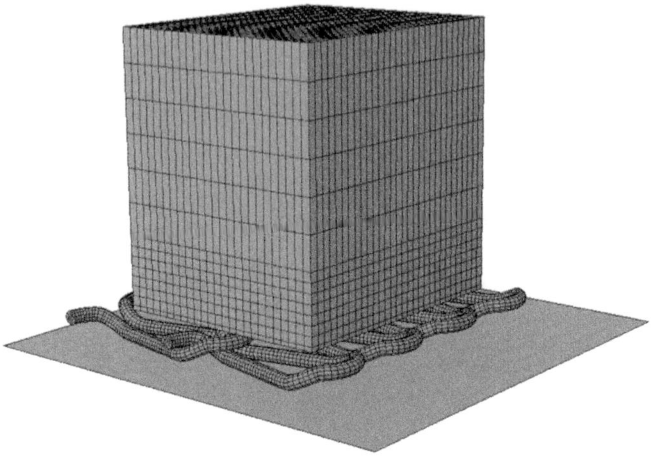

FIGURE 5.28
Model of a skin touching a knit structure. © Dassault Systèmes, a French "société européenne" (Versailles Commercial Register # B 322 306 440), or its subsidiaries in the U.S. and/or other countries.

Index